Rifle & Drill

Rifle & Drill

The Enfield Rifle Musket, 1853 and the Drill of
the British Soldier of the Mid-Victorian Period

A Companion to the New Rifle
Musket

A Practical Guide to Squad and
Setting-up Drill

S. Bertram Browne

LEONAUR

Rifle & Drill
The Enfield Rifle Musket, 1853 and the
Drill of the British Soldier of the Mid-Victorian Period
A Companion to the New Rifle Musket
and
A Practical Guide to Squad and Setting-up Drill
by S. Bertram Browne

First published under the titles

A Companion to the New Rifle Musket
and
A Practical Guide to Squad and Setting-up Drill

Leonaur is an imprint of Oakpast Ltd

Copyright in this form © 2011 Oakpast Ltd

ISBN: 978-0-85706-633-6 (hardcover)
ISBN:978-0-85706-634-3 (softcover)

http://www.leonaur.com

Publisher's Notes

The opinions of the authors represent a view of events in which he
was a participant related from his own perspective,
as such the text is relevant as an historical document.

The views expressed in this book are not necessarily
those of the publisher.

Contents

A Companion to the New Rifle Musket

S. Bertram Browne

Contents

Preface to Second Edition

When the "Enfield Rifle Musket" was first introduced into the British Army, it was apparent that some clear and definite rules should be laid down for the soldier's guidance in the treatment and management of that arm, so that, by proper care and attention on his part, it should be kept in a state of perfection, which would always ensure a firm reliance on its efficiency when required for use.

To meet this want, I was induced by Major, now Colonel, A. Lane Fox, Grenadier Guards, then Chief Instructor of Musketry at Hythe, to write my *Companion to the Rifle Musket*. Having done so, and obtained his approval of my labours, the first edition of this work was accordingly published.

The favourable reception which it met with, its ready sale and introduction into the military libraries and reading rooms, is a proof of its having accomplished the object for which it was written.

To Major General C. C. Hay, the Commandant of the School of Musketry, Colonel E. Wilford, the Chief Instructor, and Captain J. McKay, the Deputy Assistant Adjutant General of the school, I am deeply indebted for their good opinion of my labours, and their kindness, in receiving the little work on its first appearance. I here offer them the expression of my grateful thanks.

The second edition has been arranged to meet the alterations made by late improvements in the system of instruction; and such information is added, as I hope will induce soldiers to take a real interest in the beautiful weapon intrusted to their charge.

The best shots in a regiment are those men who make themselves masters of the theory and practice of rifle firing, and take the greatest pleasure in keeping their arms in a clean and perfect state. In ancient days a soldier's arms were his pride, and wealthy chieftains delighted in armour of great price. But what comparison would the arms of the ancients, with all their costliness bear to the present effective English Rifles?

When Glaucus exchanges his arms with Diomed as a token of friendship, their value is thus described by "Homer:"—

For Diomed's brass arms, of mean device,
For which nine oxen paid (a vulgar price),
He gave his own of gold divinely wrought;
A hundred beeves the shining purchase bought.

Among all savage nations their rude weapons, whether intended for war or the chase, are regarded with almost personal affection and veneration.

Surely then, it becomes the duty of every British Soldier, to hold in high esteem one of the most perfect weapons (if properly used), that science has yet placed in his keeping.

S. B. B.

Garrison Library, Chatham.

Introduction

The object of this book is, to make the soldier acquainted with an easy and effectual mode of cleaning his rifle, and of keeping it in a proper state so that it may always be ready for immediate use. Some further information is given, which, if properly attended to, will enable the soldier to become perfectly familiar with his weapon, and with its various parts; and, this being accomplished, he may always feel sure that, in his rifle, he possesses not only an intimate acquaintance, but a trusty friend in time of need.

The soldier should always be prepared to see danger without surprise, and to meet it without doubt or hesitation—hence the necessity of keeping in perfectly good order the chief arm of his defence as well as of attack.

A Frenchman once wrote a book about the best way of blowing out a candle.

There is, (said he), a right way, and a wrong one, and it often happens that we fall into the wrong one, because we have not taken the trouble of making ourselves well acquainted with the object we have to deal with.

How important is it then, that, with the introduction into the service of the improved rifle, a few easy and defined rules for its treatment—the result of practical experience—should be placed before the soldier whose duty it is, and whose pleasure it should be, always to have in the highest state of efficiency the weapon entrusted to him.

It is not alone necessary that the soldier should be taught merely how to clean and to keep his arms in order. He should further be made thoroughly to understand what particular work each separate part of his rifle has to perform in connection with the whole, and of what importance it is that not the slightest irregularity should exist in even the minutest part of the weapon. Like the bundle of sticks we read of in the fable, the rifle must be kept well together;—each part must be thoroughly clean, properly placed, and the whole screwed home. The rifle will thus be made firm, the parts will act smoothly together, and the weapon will do its work well. But if from indifference, from bungling, or from neglect, the parts are wrongly put together, or one screw left loose, the result will be failure, mischief, and disgrace.

A soldier who does not fully appreciate the trust reposed in him when a valuable rifle is placed in his hands, and who cannot be trusted to keep it clean and in proper order, will do no credit to his regiment, and be of little use to the service. The new rifle musket is made of the best and strongest materials known in the gun trade, and in the most scientific manner: it only requires proper treatment and handling to keep it in a high state of efficiency.

The present rifle musket is a very superior weapon to the old one or to the Minie Rifle. The barrel of the Enfield Rifle is fastened to the stock by a breech nail, and is also encircled by three iron bands of great strength; these bands being kept in their places by springs. Some rifles have also the "*Screw Band*," *i.e.* instead of having a spring to keep the band in its place, the band itself is fitted with a small screw, or nut, for the purpose of either loosening the band or of binding it to the barrel.

The greatest facility, however, is afforded by either of these two methods in taking the barrel from the stock; in the former by merely pressing the springs and lifting the bands, and in the latter by unscrewing the nut, and thereby loosening the bands from the barrel, so that they may be easily removed. In the old pattern musket the number of fittings were sixteen, which in the present rifle have been replaced by four, as above stated.

This system of fitting the barrel to the stock has been found to have a greater advantage than loops and pins; as in a trial to ascertain the comparative strength of the two systems, the bands were found capable of sustaining a pressure of 106 lbs. more than the old fastening with loops and pins; the breaking weight in each case being—for the old pattern, 175 lbs.; for the new, 281; difference in favour of the bands, 106 lbs.

The present bayonet and fittings differ entirely from the old ones, both in lightness and construction. First, they contain less metal: secondly, the bayonet has three grooves, which add to its efficiency; the socket is small, and fastens to the muzzle by a "*locking ring*," instead of the spring in the stock. In order to enable the soldier to appreciate the advantage of this seemingly trifling, but, in reality, important improvement,—the "locking ring,"—he should be informed that the cavity made in the stock of the old muskets to receive the bayonet spring, was a source of great injury to the gun, inasmuch as dirt and wet would enter, and thereby cause rust to accumulate about the barrel and the stock, which could only be removed by dismounting the barrel.

The ramrod [1] is constructed with a swell near the head, which also acts as a spring to keep it in its place. The instructor should shew the soldier that the swell of the ramrod will greatly assist him (when the bullet requires more force than usual through fouling in the barrel), in starting the bullet a few inches down the barrel, by the facility it offers for grasping the rod at that particular point. The head of the ramrod is made in the form of a jagg, so that instead of having a separate jagg, carried in the pouch, it now forms part of the ramrod itself.

The superiority of the swivel lock over the hook lock of the old musket arises mainly from the principle on which the tumbler is acted upon by the main-spring. In the present swivel lock, the main-spring is connected with the tumbler by means of a chain or swivel, which is suspended from the shaft of the tumbler to receive the claws of the main-spring, so that, when the tumbler revolves, no impediment arises from friction. In the swivel lock friction is reduced so much that it never can exceed one *per cent.*; consequently as soon as the sear nose is disengaged from the bent, the tumbler is put in motion without offering in itself any opposition by friction.

The beautiful principles developed in the construction of a good swivel lock cannot be surpassed. The formation of the tumbler, when assisted by the swivel, gives an arrangement of leverage partaking of the multiplicate; for the weight, when approaching full cock, is lessened by the lever bringing the moving force into the immediate vicinity of the tumbler, and when down on the nipple increasing or multiplying that force by the divergence from the axle.

Mr. Greener, a celebrated gunmaker, says there is a great degree of skill displayed in the making of locks, though to the casual observer it does not appear. On the simple hanging of the swivel depends all the sweetness of the play of the main-spring, and on the placing the hole for the sear pin, depends the sweetness of the sear playing on the tumbler.

1. The Rifles fitted with "screw bands" have a straight rod.

15

All locks for percussion should have the greatest strength of main-spring at the moment they strike the nipple, or what is termed when the lock is down. On the pitching the sear depends the cutting of the bents, and on their formation the danger of the lock catching at half cock, when the trigger is made to pull easy.

In the hook lock (*vide* **plate 11**) the end (or toe), of the main-spring must travel up an incline when cocking, and down when firing; consequently a large amount of leverage is destroyed by friction, which is very considerable. Hence the velocity and consequent force of the hammer in the swivel lock is greater, and its action more free and smoother than that of the hook lock.

Two springs of equal force and flexibility, fitted to locks of the hook and swivel pattern, will be found to communicate velocities varying according to the amount of friction generated in their action.

It is almost superfluous to observe, that with such an admirable weapon placed in the hands of the soldier, he is bound in honour to keep it in a state of perfect efficiency, and ready for use at a moment's notice.

Too much care, therefore, cannot be taken to ascertain its condition, by frequent inspections, not only on parade, but also in the barrack rooms. It must be perfectly free from rust, and other damage. It cannot be too often impressed upon the mind of the soldier, that if rust be allowed to accumulate in the barrel the bullet will be prevented from taking the rotatory motion, which is actually essential to the accuracy of its flight. Hence it is proved that the accuracy of the flight, and, consequently, the due execution of the bullet, is not dependent solely on the rules laid down in the theory and practice of firing, but that the perfect condition of the arm itself is necessary to secure due and efficient execution.

Defects in the structure of the rifle itself will sometimes occur; and these, the soldier, if ignorant or inexperienced in its treatment, may be unable to detect or rectify. He may never hit the mark; he may get the name of a bad shot, and his bad firing be attributed to his incompetence; whereas, if there were no defects in his rifle, he might prove to be as good a shot as his comrades. This should be thoroughly explained to the soldier, in order that he may become the more zealous in his efforts to detect and rectify any accidental defects in his weapon.

The rifle may not have a proper bore; such a defect is, however, of very rare occurrence. In such cases the cartridge is found to ram down hard, or on the contrary, is very loose in the barrel. Such faults should be immediately reported.

Many important remarks on the inherent defects of rifles will be found in the various valuable works and lectures on rifle firing.

Soldiers, more particularly the younger ones, should be taught, when cleaning their arms, to take the greatest care not to rub or damage the *foresight*; this many unthinkingly do, not knowing how greatly such damage may operate against them at the time of trial. The lock, too, must not only be quite clean, but oil must also be applied to the parts that rub. For want of a little oil, the trigger will sometimes pull hard, and cause the firer to alter the direction of the piece while in the act of firing. This affords further proof of the importance of keeping the lock in good easy working order.

Another important point cannot be too strongly enforced, or too strictly attended to. If, through accident, carelessness, or neglect, any dent in the barrel should be made, or the barrel itself become bent, it may burst in the soldier's hands. If, therefore, the lock should become wood-bound, the nipple injured, or any part of the rifle damaged, which it is not in the soldier's power to remedy, he must immediately report such damage, that the armourer alone may repair it. On no account must the soldier himself attempt such repairs.

Young soldiers are sometimes not aware that the explosion of a cap upon the nipple, when there is no charge in the barrel, causes more rust than the firing of a charge. Instructors must, therefore, insist upon the necessity of thoroughly cleaning and oiling the rifle after a practice of snapping caps.

In order to keep the muzzle and lock as free as possible from exposure to rain, the rifle must be carried at the "*Secure;*" the rain will then run off the muzzle, while the lock is safe under the arm.

When "*Ordering Arms,*" the butt should be placed gently on the ground, thereby avoiding injury to the mechanism of the lock, which will occur if the butt is struck violently against any hard substance.

The soldier having learned the names of every part, and seen the manner in which the rifle is taken to pieces, the instructor should then make him dismount the arm himself, and for each thing that he is about to perform, repeat aloud the instructions for doing it; by this means he soon learns how to give instructions to others.

The preceding remarks are not intended as mere introductory matter, to be forgotten as soon as read, but as a part of the instruction which should be read and carefully explained by the instructor to his pupils. He should also explain to them (when opportunities offer), the vast and important modifications guns and gun locks have undergone since they were first brought into use. How great the difference between the first method, *viz.*—ignition from a match held in the hand—the second, by that of a match placed in the jaws of a cock, and brought into contact with the priming, by means of a spring trigger: the third, by the rotation of flint and steel from the wheel lock; or the fourth, by the flint lock, and the smooth and instantaneous production of fire by the present well-finished percussion or detonating lock.

The first spring locks, it is said, were made in Germany and in the city of Nuremberg, in 1517. Could the early gun-lock makers witness the perfection to which the present locks have been brought, we may imagine with what amazement and admiration they would regard them.

It is hoped that these remarks may be the means of inciting the young soldier to seek for further information respecting the use of military weapons, and the great improvements they have undergone since their first invention by man.

The present system of rifle instruction tends in the greatest degree to improve the mind of the soldier, causing him to study and master (during his leisure hours), such points of instruction as in the lecture-room often seems difficult to understand. "I know," (says Major General Hay, the commandant of the School of Musketry), "of no system or subject of training that has been introduced into the Infantry Service, better calculated to promote the interest of the school than that of musketry instruction, which calls into exercise the thinking faculties of the men, and incites in them a desire to acquire the art of reading, and the knowledge of other useful subjects, that they may understand this branch of their military training more perfectly. I have witnessed the greatest desire on the part of the men for opportunities for mental improvement, and am convinced that the system under consideration, supported and stimulated by the recent scheme for prizes for good shooting, will do much to increase the attendance at school."

A good shot will have his rifle always in a clean and perfect state.

It has been by attention and study that British Soldiers are brought to a degree of perfection in the use and knowledge of their arms, that enables them to surpass all troops in the world, and to remain unrivalled among the armies of the earth.

In conclusion, I would earnestly urge upon rifle instructors, to point out to their pupils how to take advantage of the many excellent works contained in the military libraries; from which they may gather the whole history of firearms, from the first invention of gunpowder, through the successive steps by which the rifle musket has attained its present efficiency. Every encouragement is now held out to those soldiers who perseveringly apply their mind to the attainment of excellence in their profession as marksmen. They have opportunities of improving themselves by the schools and libraries, that many thousands of hardworking men in civil life yearn for, but have not the privilege of enjoying. It is in the power of every young soldier to store his mind with such an amount of sound experimental knowledge, as shall enable him to say with truth and pleasure,

Of highest genius 'tis my pride,
To comprehend what art has done;
To know the law her steps that guide,
And share the glories she has won.

S. B. B.

Description of Plates

PLATE 1

Description of Enfield Rifle Musket, 1853,
(fitted with solid bands.)

Musket—
    Length without Bayonet, 4 ft. 7 in.
    Do.   with   do.   6 ft. ½ in.
    Weight without do.   8 lb. 8 oz.
    Do.   with   do.   9 lb. 3 oz.

Barrel—
    Weight, 4 lb. 2 oz.
    Length, 3 ft. 3 in.
    Number of Grooves, 3.
    Depth of   do.   ·014 in.
    Width of   do.   ·262 in.
    Spiral, ½ turn in 3 ft. 3 in.
    Bore $\left\{ \begin{array}{l} \text{Diameter at Breech} \\ \text{Do.} \qquad \text{at Muzzle} \end{array} \right\}$ ·577 in.

Lock—Swivel.
    Main Spring draws at half cock, 13 to 14 lb.
    Pull of the Lock, 13 to 14 lb.
    Sear Spring draws 7 to 8 lb.
    Trigger drawers, 7 to 8 lb.

Charge—
    Bullet—Weight, 530 grs., length, 1·03 in.,
           diameter, ·568 in., windage, ·009.
    Powder, 2½ drams FG (fine grain).
    Weight of 60 Rounds and 75 Caps, 5 lb.
        8 oz. 4 drs.
    Charge of Powder in pro- $\big\}$ Bullet weighs
        portion to Bullet   7·75 Charges.

PLATE 2

THE SWIVEL LOCK (OUTSIDE).

A Sear Spring Pin
B Sear Spring Pin
C Bridle Pin

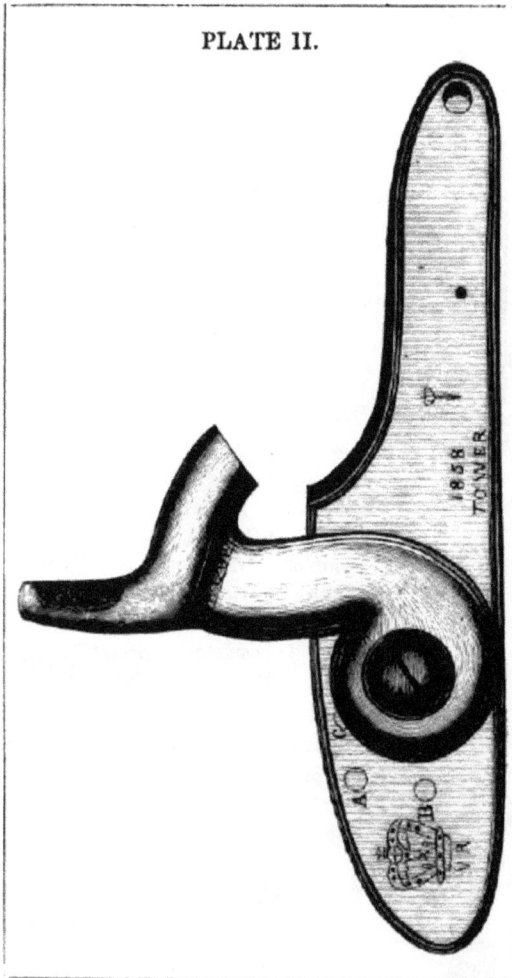

PLATE II.

PLATE 3

THE LOCK.

Name the limbs of the Lock in the order in which they are removed.[1]

A Main Spring   E Hammer
B Sear Spring   F Tumbler
C Sear   G Swivel
D Bridle   H Lock Plate

PINS.

a Tumbler   c Sear
b Sear Spring   d Bridle

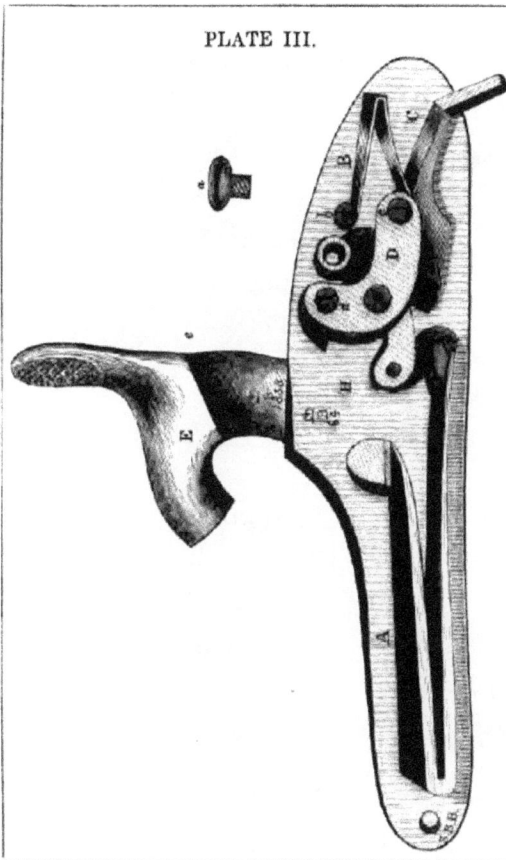

PLATE III.

---

1. *Vide* Order of Instruction. (First Lesson).

PLATE 4

THE STOCK,

A The Nose Cap	B The Bands
C The Swell	D The Lock Side
e Projection	F The Head
G The Small	H The Trigger Guard

i The Trigger Plate;—(*Vide* Plate 5, fig. 2.)

k The Trigger—(*Vide* Plate 5, fig. 2.)

L The Butt—*a* The Toe *l* The Heel

M The Heel Plate	N The Band Springs
O The Breech Nail	P The Side Nails

THE BARREL.

A The Muzzle	B The Front or Fore Sight
C The Back Sight	D The Nipple Lump

THE BAYONET.

A The Blade	B The Socket
C The Locking Ring	

PLATE 5

Fig. 1.—Back or Elevating Sight.

a The Bed	*e* The Pin
b The Flanges	*f* The Spring
c The Slider	*g* The Cap
d The Flap	

Fig. 2.—The Trigger.

a The Trigger	*d* The Stud
b The Finger	*e* The Plate
c The Box	

PLATE 6

Fig. 1.—The Breech Pin,

a The Face	*d* The Tang
b The Screw	*e* Breech Nail Hole
c The Neck	

Fig. 2.—The Nipple.

a The Cone	*d* The Screw

PLATE 4

PLATE IV.

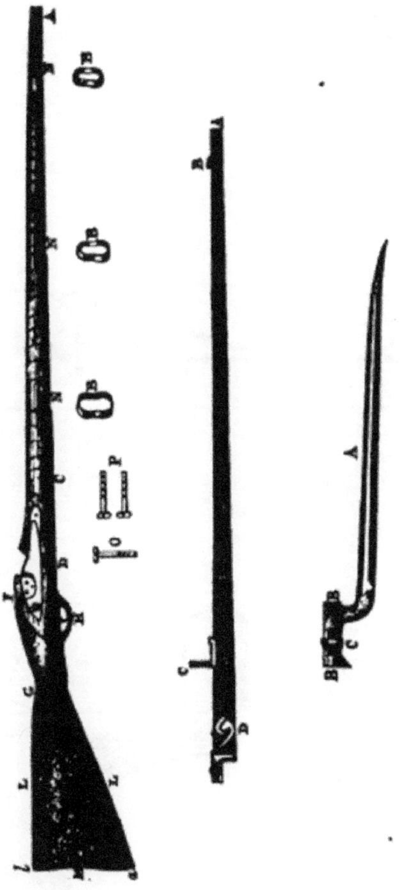

PLATE 5

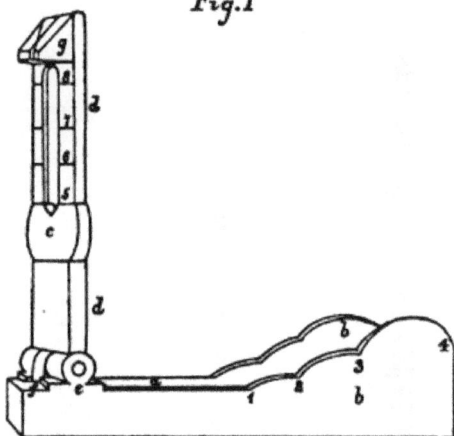

PLATE V.

Fig.1

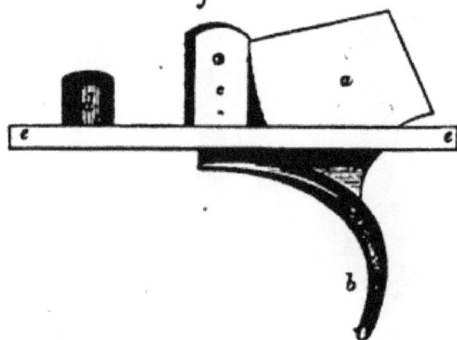

Fig.2.

PLATE 6

PLATE VI.

Fig. I

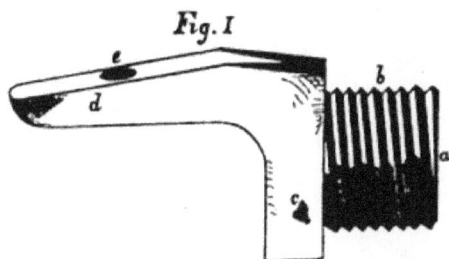

Fig. 3

Fig. 2.

b The Square *e* The Touch Hole (*vide* Section of Nipple *f*)
c The Shoulder

The Ramrod (in part)
a The Head and Jagg *b* The Swell[2]

PLATE 7

Name the various parts of the limbs of the Lock[3]

Figs. 1 and 2.—Main Spring (two ways).

A Catch   D Bend
B Return   E Claws
C Stud   F The Spring

Figs. 3 and 4.—Sear Spring (two ways).

A The Eye   D Bend
B Return   E Spring
C Stud   F Toe

Fig. 5—The Sear.

A Arm   D Neck
B Body   E Nose
C Eye

PLATE 8

Fig. 1.—The Bridle.

A Stud   D Tumbler Pivot Hole
B Foot   E Sear Pin Hole
C Bridle Pin Hole

Fig. 2.—The Hammer.

A Mouth   D Neck
B Head   E Body
C Comb   F Tumbler Square Hole

Fig. 3.—The Tumbler.

A The Pivot   E Half Bent   B Bearer
F Full Bent   C Shaft   G Axle
D Swivel Stud or Pivot Hole   H Squares
I Pin Hole

---

2. All Rifle Muskets fitted with the solid bands and springs have this description of "Rod." Those fitted with the "Screw Bands" are straight from "head to toe."
3. *Vide* Order of Instructions. (Second Lesson).

PLATE 7

PLATE VII.

Fig. I.

Fig. 2.

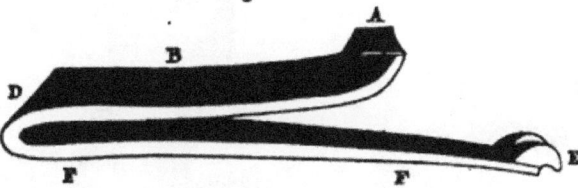

Fig. 3.

Fig. 4.

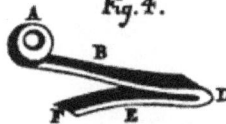

Fig. 5.

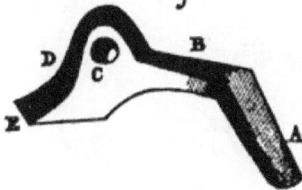

PLATE 8

PLATE VIII.

Fig.2.

Fig.1.

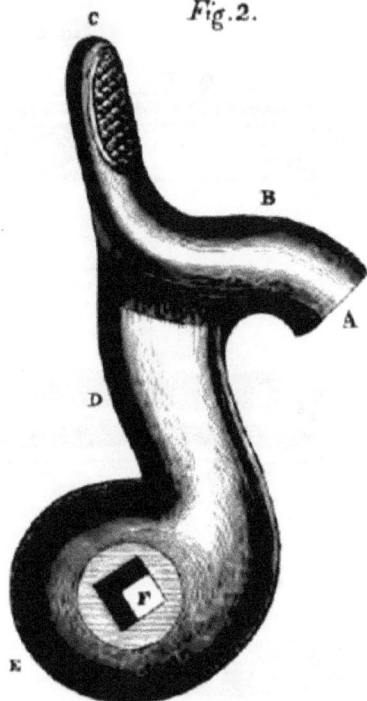

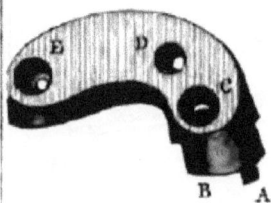

Fig.3.

Fig.4.

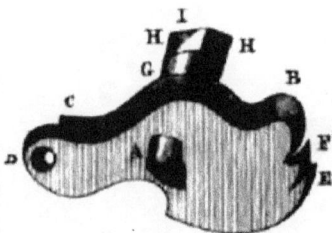

Fig. 4.—The Swivel.

A Body   B Studs or Pivots

PLATE 9.

The Lock Plate (inside).

A Front Side Nail Hole
B Main Spring Stud Hole
C Fore Stud
D Bridle Stud Hole
E Bridle Pin Hole
F Hind Stud (and Hind Side Nail Hole)
G Sear Spring Pin Hole
H Sear Spring Stud Hole
I Sear Pin Hole
K Tumbler Axle Hole

N.B.—For Third Lesson, *Vide* page 60.

PLATE 10

Improved Nipple Wrench and Spring Cramp.

Fig. 1.

Main Spring in Cramp as taken from the Lock in dismounting.—
(*Vide* First Lesson.)

Fig. 2.

A The Nipple Wrench
B The Cramp Hook
C The Worm or Double Wrench
D The Turnscrew,
E The Oiling Wire
F The Reservoir for Oil
G The Lever
H The Stud
I The Pricker [Fig. 1., I.)

Some Nipple Wrenches have been fitted with a "Ball Drawer" (Fig. 3), and a "Drift" (Fig. 4), both used for extracting the bullet from the barrel when necessary. The former screws into Nipple Wrench; the latter into the reservoir, and acts for the same purpose as the oiling wire.

PLATE 9

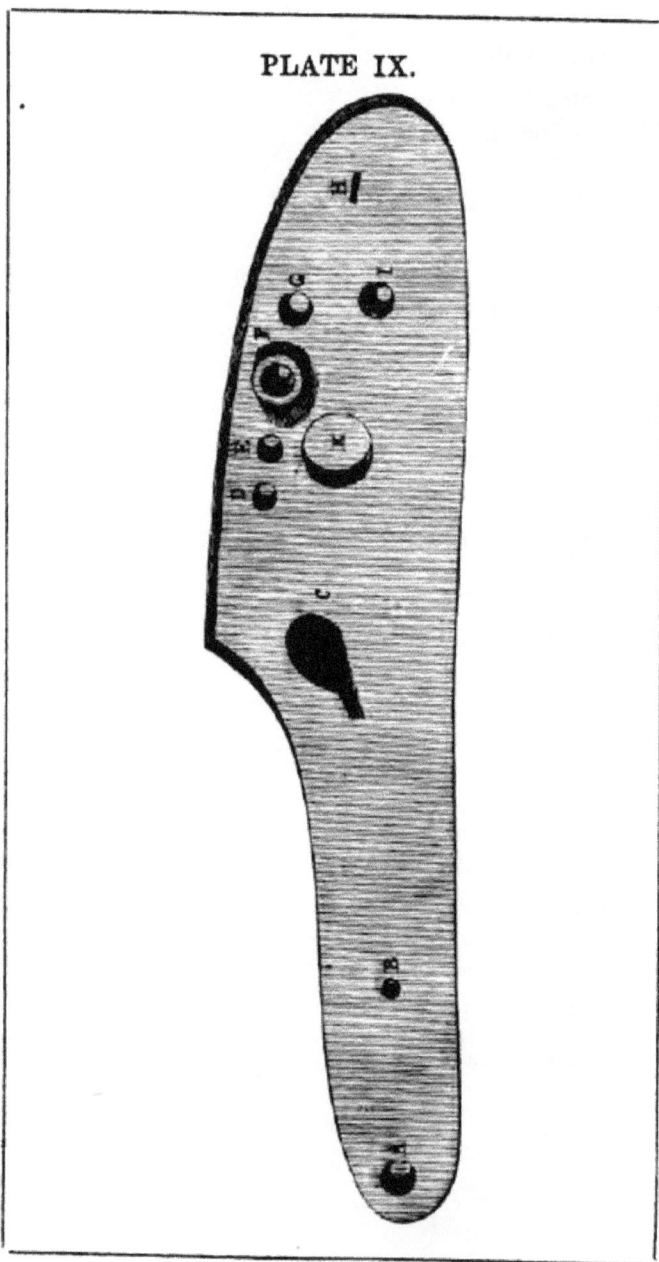

PLATE IX.

PLATE 10

PLATE X.

Fig. 3.

Fig. 2.
A

FigI

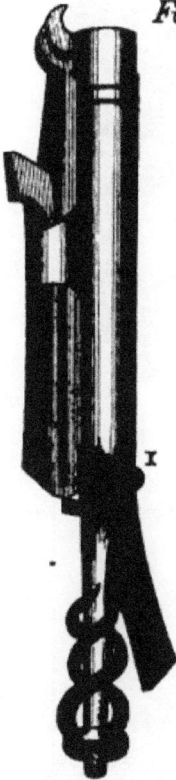

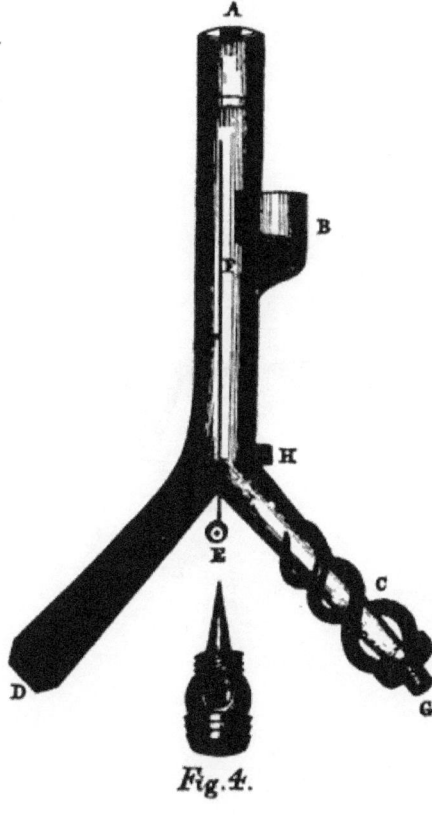

B

H

I

E

C

D

G

Fig. 4.

PLATE 11

The Hook Lock.

A The Main Spring
a The Eye of the Main Spring
B The Hook of the Tumbler
C The Bridle
D The Sear
E The Sear Spring
b The Bridle Pin
c The Sear Pin
d The Hind Stud

PLATE XI.

The Circular Trigger

As the circular trigger has been fitted to several of the rifle muskets, I think it very essential that the Soldier should have an opportunity of thoroughly understanding the excellent principles upon which this new invention acts. The following information will therefore prove of great service to those who are unacquainted with its merits, and enable them in practice to prove the very great advantage gained by understanding the theory of the invention.

From my own experience as a teacher at the School of Musketry at Hythe, I always found my pupils most anxious to possess a rifle fitted with the "Circular Trigger;" they were always certain of making better firing than with those fitted with the Lever Trigger, **Plate 5.**, fig. 2, and the result of the practice generally proved the correctness of their judgement.

THE CIRCULAR TRIGGER
Invented by Captain Harris, Royal Marines Light Infantry.
PATENT.

The object of this invention is to do away with a very great and universally acknowledged defect in the "pull off" of firearms fitted with the present or lever trigger.

The well-known difficulty at present experienced in obtaining a *Fine, Light, and Safe* pull off, arises from the inability of a simple *Lever* to combine *all* these qualities, for it is well known and admitted by all Sportsmen and Gunmakers, that with a *safe* or *strong* lock, if a *short* pull is required, it is so *hard* as to *completely destroy* the best aiming, and a *long* pull is such a *series* of *jolts* as to be quite inadmissible. Again, if the lock is made *light*, a *loss* of *safety* ensues to a certainty.

Thus, instead of the "*Fineness*," "*Lightness*," and "*Safety*" we seek, we find ourselves hemmed in by "Hardness," a jolting pull or "Drag," or a *Loss* of "Safety;" and we cannot avoid one of these defects without bordering on another of them.

The result has been, that the hardness and drag being so extremely disagreeable, a loss of safety has been submitted to for the purpose of obtaining the only kind of pull tolerable in a lever, *i. e.*, fine and light, or *short* and *light;*—thus, the *Defect* of the *Principle used*, and not the *Requirements of Aiming*, have made a *short* and *light* pull the *proverbial demand*, to the detriment of *safety;*—that this deduction is true is proved by the existence of the hair trigger, which is the *perfection* of *shortness* and *lightness* of pull; but it is so *dangerous* as to be quite shunned for general service and usefulness.

Now, the circular trigger is a cam, or union of the *Lever* with the *wedge*, which gives such enormous power that a strong or safe lock may be discharged with a very light and easy pull, and by the particular application of the mechanical principles used in its construction, the pull is also *even throughout*, which entirely obviates the *hardness* or *jerk* and "Drag" so detrimental to good shooting.

Thus *Safety* and *Lightness* of pull, two of the requirements mentioned above, have been obtained by the use of the circular trigger, the third point *"shortness"* is not attained *in the Trigger*, nor could it be without springs and complications which have been shown in the hair trigger to make it dangerous. But shortness of pull is attainable, *if* and *when* required, in the circular trigger, because the force requisite to use it is so very little (about 3 lbs. for a military gun, and 1 for a sporting gun), that the *Finger can make the action of pulling as short as necessary*, and after some practice with this trigger, by pulling it partly off and there resting, the final discharge is made, not only as fine and light as the best hair trigger, but perfectly safe, and quite *controllable*, whilst one is almost afraid to touch the hair trigger. The extremely easy and *sliding* pull of the circular trigger makes its use not only very *pleasant but greatly assists steadiness of aim*.

To sportsmen this trigger offers the advantage of safety when carrying the gun at full-cock, for no inanimate object is at all likely to retain its hold sufficiently to pull the trigger the whole way off, and when it quits its hold, the sear is made to return to its place in the top bent.

Time is saved in making use of a good rifle aim when obtained, for it is perfectly safe to put the finger already on the circular trigger when aiming, and there is nothing to do but to shut it up close in the twinkling of an eye when the aim is complete, whereas one dare not put his finger on the hair or common trigger till *after* the aim is perfect, and in that moment it may be lost.

Figure 1 shows the principle of the circular trigger, *a d f*, or *c d f* being the lever part, and *c b a* the wedge, which is struck from a particular centre *o*; *d* is the axis of the lever and trigger. As the trigger, revolving upon the axis *d*, moves the wedge *a b c* in the direction of the arrow, the sear resting on the surface of the wedge is gradually and evenly forced up until the gun is discharged.

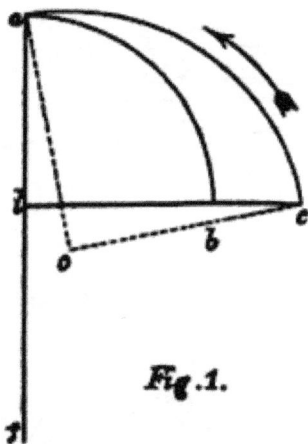

Fig.1.

Figures 2, 3, and 4, show the form and action of the circular trigger—*i. e.*, Fig. 2 shows the sear securely locked in the safety-notch of the trigger at half-cock, making a pull-off in this position, or any danger, impossible. When the cock is down the sear is similarly locked. Fig. 3 shows the position of the sear and trigger at full-cock, the former having escaped from the safety-notch in the act of cocking-the trigger is now free to move. Fig. 4 shows the positions of the sear and trigger at the moment of discharge, the sear having been wedged up as before explained.

The circular trigger is readily applicable to all kinds of firearms, and requires but a slight and easy alteration in the gaiting of the tumbler, and the form of the sear arm.

Besides the great advantage of removing the jerk and drag, and admitting the use of a safe and strong lock, a careful inspection of this principle will show, that—

1. It simplifies the form of the lock about the half-bent.

2. Ensures greater safety.

3. Saves the lock a deal of wear and tear.

PLATE 12

PLATE XII.

Fig. 2.

Fig. 3.

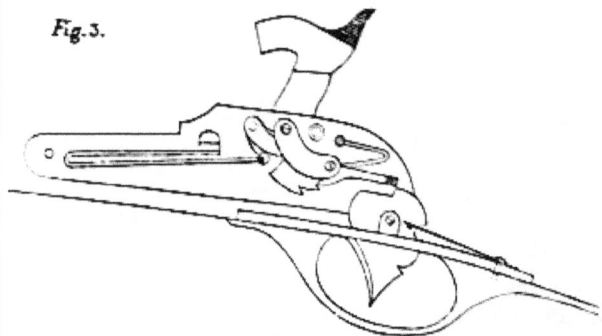

Fig. 4.

4. Keeps the wet and dirt out of it.

5. Cannot get out of order, being so extremely simple in formation.

6. Is inexpensive; for, in new guns it would cost only a slight addition to the expense, and not much to chose in use.

7. It allows of the gun being *safely* carried at *full cock*, or even the half bent being quite dispensed with in sporting guns.

8. It reduces the discharge of the gun to *one single motion* when the aim is ready, the contraction of the forefinger, and this in the *time* most agreeable to the person using it.

A continued and lengthened trial has been made of this invention by the Government: it has been pronounced, by competent authority to be sound in principle, safe and unobjectionable in practice, and to overcome that defect which in the old trigger cannot be obviated by *any amount* of practice.

A Regiment of Infantry is now being armed with triggers on this improved and important principle.

January, 1858.

B. De La Cour, 327, High Street, Chatham, is licensed, under the Patent, to apply the above most important invention to every description of firearms.

Plate 13

PLATE XIII.

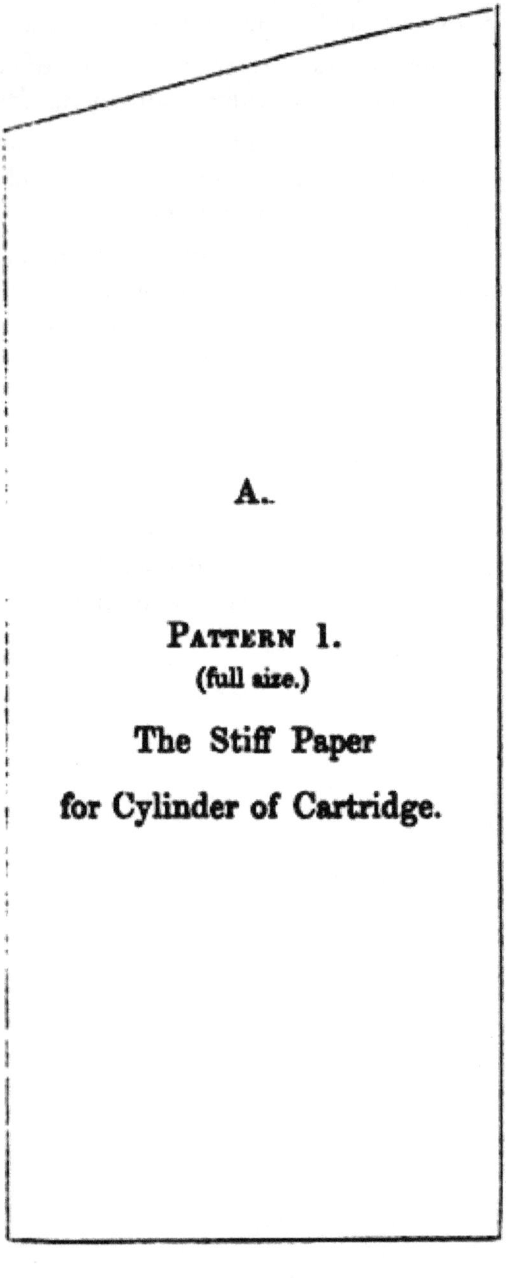

A.

Pattern 1.
(full size.)

The Stiff Paper

for Cylinder of Cartridge.

PLATE 14

PLATE XIV.

B

PATTERN 2,

(full size,)

The Iuner Envelope.

Plate 15

PLATE XV.

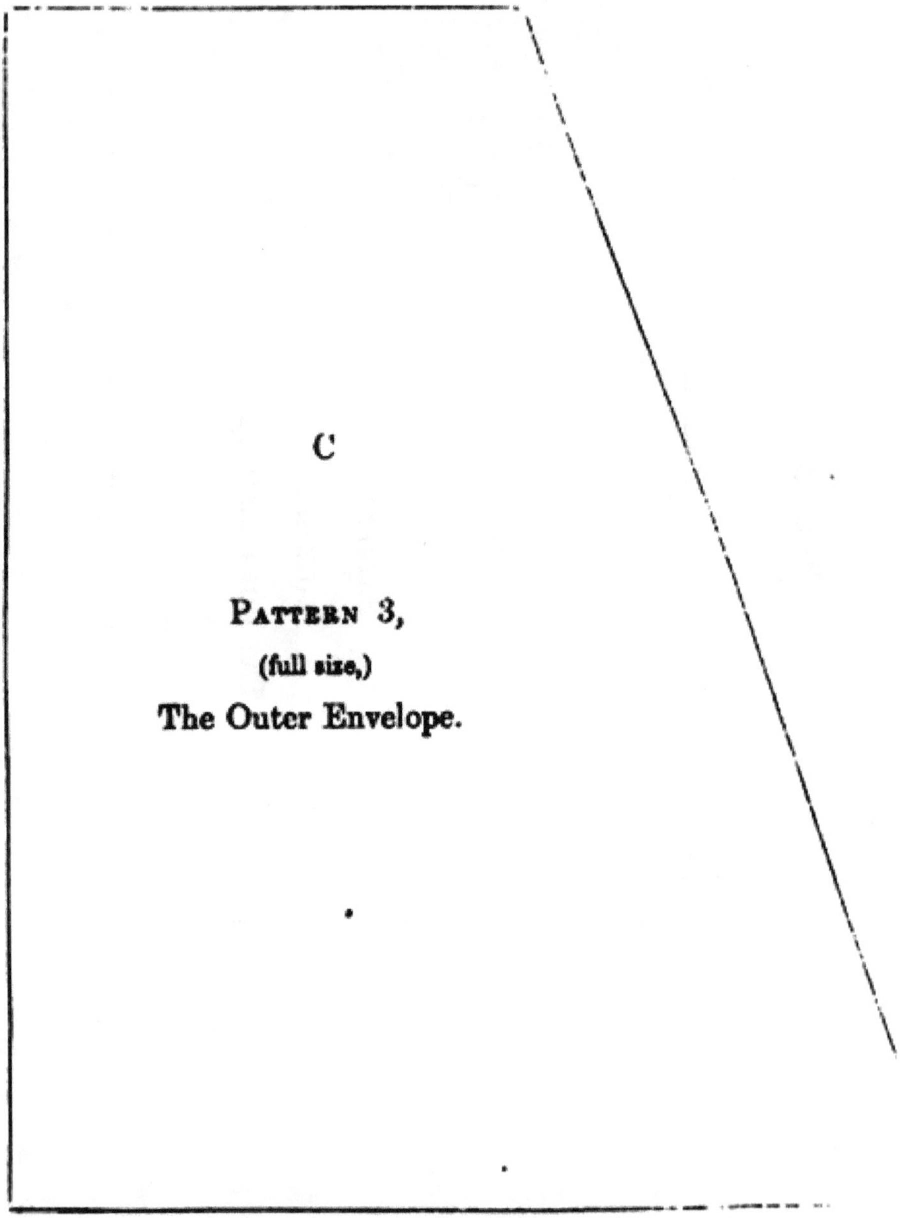

C

Pattern 3,
(full size,)
The Outer Envelope.

PLATE 16

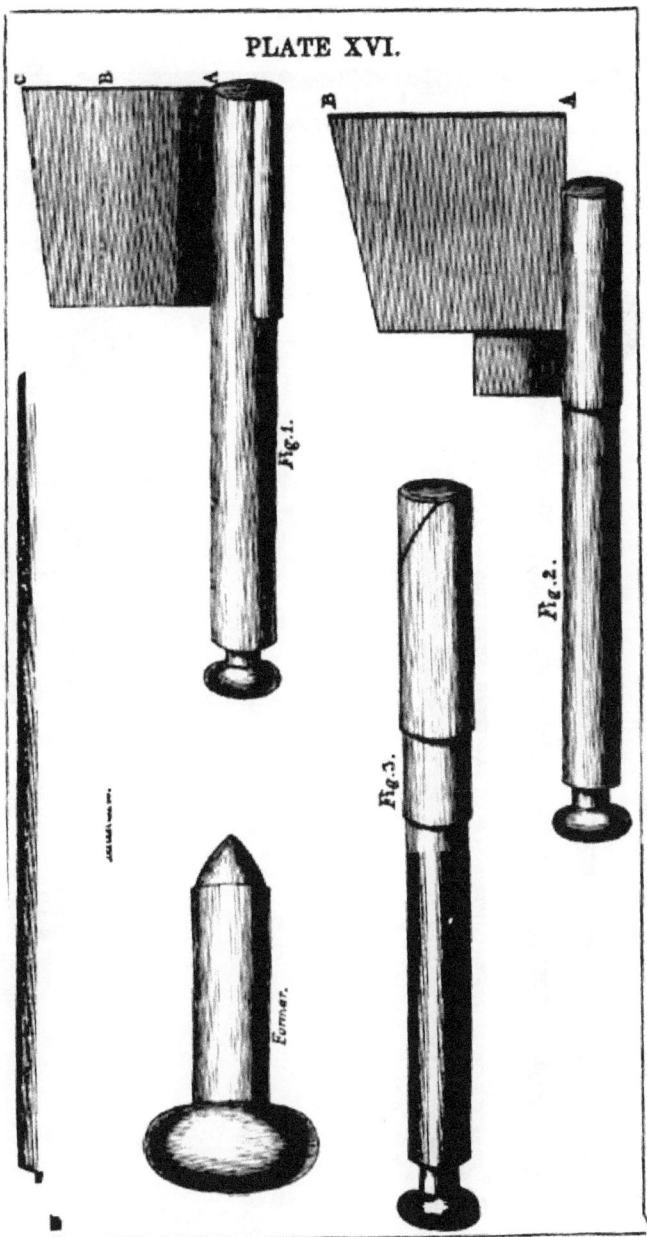

PLATE XVI.

Fig.1.

Fig.2.

Fig.3.

Former.

PLATE 17

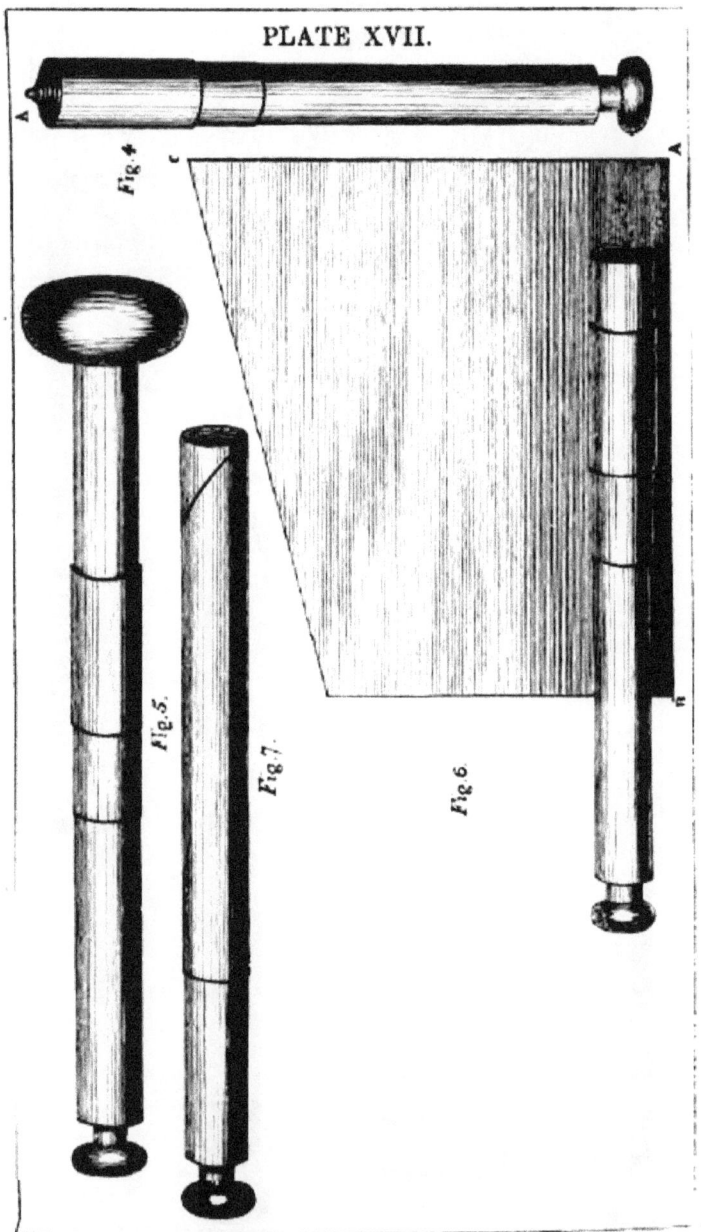

PLATE XVII.

Fig. 4

Fig. 5

Fig. 7

Fig. 6

PLATE 18

PLATE XVIII.

Fig.8.   Fig.9.   Fig.10.

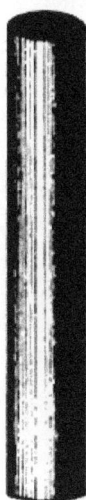

B

A

Lessons

CLEANING ARMS

Every soldier is to be taught the names of the different parts of his rifle, and thoroughly acquainted with the rules for cleaning and keeping them in proper order. This is the first exercise in which the soldier is to be instructed.—*Vide Queen's Regulations.*

Instruction under the head of "Cleaning Arms," *Vide* "Instructions of Musketry," should be conveyed to the soldier in eight lessons, as follows, *viz.*:—

First Lesson.—Name the limbs of the lock, and the other principal parts of the rifle, as also how to dismount the lock.

Second Lesson.—Names of the various parts of the limbs of the lock.

Third Lesson.—How to clean the lock and rifle, and to keep them in a serviceable condition.

Fourth Lesson.—Remount the lock.

Fifth Lesson.—Explanation of the uses of the several parts of the lock, in two lessons.

Sixth Lesson.—Continuation of the fifth lesson.

Seventh Lesson.—How to guard against the rifle and lock getting out of order, in two lessons.

Eighth Lesson.—Continuation of the seventh lesson, showing strongly the necessity of preserving the rifle from the slightest injury, and a careful preservation of the ammunition.

FIRST LESSON.
Directions for Dismounting the Lock.

It will seldom be found necessary to take the lock to pieces if proper care and attention has been taken of the arm.

1st. Take out the tumbler pin.

2nd. Put the lock at full cock, and take off the main spring. To do this, take the cramp (**Plate 10**. fig. 1), and remove the screw drawer from the nipple wrench; then place the hook of the cramp on the return of the main spring, press the cramp well to the lock plate. See that the stud of the cramp is close to the bend of the main spring. Let the hammer down so as to take it off the bearer; then lift the stud of the main spring gently out of the plate, the spring remaining in the hook of the cramp.

3rd. Take off the sear spring. To do this, unscrew two or three threads of the sear spring pin. Put the edge of the turnscrew under the bend of the spring, so as to lift the stud of the spring from the lock plate; then unscrew the pin entirely, and remove the spring.

4th. Unscrew the sear pin, and remove the sear.

5th. Unscrew the bridle pin, and remove the bridle; if the stud of the bridle should stick, tap upon the lock plate with a piece of wood, and so loosen the stud.

6th. Take off the hammer. To do this, let the body of the hammer rest fairly in the hollow of the left hand, and in the position of half cock; then give a few smart taps on the inside of the neck of the hammer, until it gradually leaves the squares of the tumbler.

7th. Take out the tumbler.

8th. Remove the swivel from the tumbler.

Whenever there are more than one stand of arms being dismounted in the same place, the soldier should place carefully the parts of his own rifle in some receptacle, which will prevent them from being lost or intermixed with other arms.

The greatest possible care and attention is required on the part of instructors, to see that their pupils dismount the lock in a neat and proper manner; all awkwardness must be banished, as it requires great dexterity in taking it to pieces, and more so in putting it together. To do it well calls not only for much mechanical skill, and a light but firm hand; but it also requires a personal interest in the task, which none but the owner will ever feel. Many good gun locks have been spoilt by a careless hard-handed fellow doing too much, and many also by some lady-handed man doing too little.

N. B. For Second Lesson, vide page 24

Third Lesson.
Instructions to Clean the Lock and Rifle, and to keep them in a serviceable condition

1st. When the lock is dismounted take every part singly, and wipe it perfectly clean with an oiled rag, and afterwards with a dry one.

2nd. If any specks of rust are seen on any of the interior parts of the lock, lock plate, and particularly the tumbler axle hole, the same must be removed by dropping a little oil on the spot, and with the point of a bit of wood, rub the rust clean out, and wipe the surface dry; no brick-dust or powder of any kind is to be used for this purpose, as it would have the effect of removing from the limbs thus treated, (and which are not steel,) the case hardening, and thereby render them more liable to rust.

3rd, To keep the lock in proper condition, observe that in remounting the lock, the threads of the several pins, as also the pivot and axle of the tumbler, and the pivots of the swivel should be touched lightly with oil, before putting them in their respective places, so that they may work easily.

4th. The other frictional parts of the lock to which it is essential to apply oil (which should be animal, and not vegetable oil), are the nose of the sear where it touches the bents, and the toe of the sear spring where it touches the sear.

In oiling the parts bear in mind, that a great quantity of oil is not required. Put only the smallest drop with a feather, or with the point of the "drift," for by putting on too much oil, it is liable to clog the parts.

To Clean the Rifle.

1st. Place the rifle at full cock, and take out the rod.

2nd. Put a piece of rag or tow into the jag and twist it round so as to cover it.

3rd. Hold the rifle in the left hand, with the forefinger and thumb in line with the muzzle, and at the full extent of the arm, the barrel downwards, the heel of the butt resting on the ground.

4th. Pour about a quarter of a pint of clean water into the barrel with care, so as to prevent any of it getting between the stock and barrel or into the lock through the tumbler axle hole, or the snap cap that covers the nipple.

5th. Put the ramrod immediately into the barrel and sponge it carefully up and down, forcing the water through the nipple to remove the dirt or fouling, as also to clear the touch hole.

6th. Wipe the barrel well out with a dry rag, or tow, until it is clean and quite dry; after which, use an oiled rag, then put the muzzle stopper in the barrel, and the snap cap on the nipple; wipe the mouth of the hammer and let it down on the snap cap.

7th. The following morning, and upon every occasion before using the rifle, wipe the barrel out clean and perfectly dry.

The fouling which settles on the stock near the nipple lump or trigger plate, should be removed with a rough rag; using water or scraping with a sharp instrument is strictly forbidden.

In order to prevent the water soaking into the stock, and at the same time to give it a smooth and polished appearance, rub it over with a little linseed oil, also apply a little bees' wax between the stock and barrel, and between the lock plate and stock; so as to prevent water getting under the barrel or into the lock.

If the interior of the barrel be allowed to become rusty, the increased resistance to the passage of the bullet will probably cause the latter to "strip" (or pass out of the grooving), or else the wood plug may be driven through the bullet, and the arm for the time rendered useless or dangerous.

Fourth Lesson.
Remount the Lock.

1st. Attach the swivel to the shaft of the tumbler,

2nd. Place the axle of the tumbler in the axle hole of the lock plate, keeping the bearer well against the hind stud.

3rd. Place the bridle on the pivot of the tumbler, with the stud properly fixed in the lock plate, and screw home the bridle pin.

4th. Place the sear between the bridle and lock plate with its nose against the tumbler, and screw home the sear pin.

5th. Take the sear spring and partly screw it to the lock plate, then, with the thumb of the left hand, *press*[1] the spring against the body of the sear, until the stud enters the stud hole, then screw home the sear spring pin.

---

1. Without this pressure, it is not easy to get the stud of the spring into the hole in the lock plate, and great care is also necessary to see that the stud sits fairly in its place.

6th. Turn the lock over in the left hand, and place the hammer on the squares of the tumbler, in the same position as if down upon the nipple.

7th. Connect the pivots of the swivel to the claws of the main spring; put on the main spring, by placing its stud in the lock plate, with the catch under the fore stud, after which draw the lock up to full cock and take off the cramp.

Feel that the parts work smoothly together, by lifting the hammer and sear up and down two or three times, then let the hammer down to ease the spring. Great care must be taken that no grit or dirt gets between the tumbler and lock plate, or the bridle and tumbler, when putting the lock together.

FIFTH LESSON (IN TWO PARTS).

Explanation of the uses of the several parts of the lock, and how they may be put out of order. (in two lessons.)

The Main Spring,—This spring acts upon the tumbler to draw the hammer down upon the nipple, the stud which is attached to the return is necessary to keep it in its place, which, together with the fore stud, which secures the catch to the lock plate, presents a resistance by which the action of the spring is ensured.

The Sear Spring.—This spring acts upon the sear; the toe of the spring bears against the body of the sear, and presses it outwards, whenever the tumbler is revolved; by drawing the hammer backwards, thereby causing the nose of the sear to enter the half and full bent; and thus keeping the cock at half or full cock as may be desired. The stud, which is attached to the return, is necessary to keep it in its place; and, together with the sear spring pin, secures it to the lock plate, thus presenting a resistance as to ensure a free action of the spring.

The Sear.—The sear acts as a stop on the movements of the tumbler by its nose entering the half or full bent; thus keeping. the lock at half or full cock. The nose of the sear is so constructed as to fit the bents exactly. The arm of the sear is that part on which the trigger acts, to raise the nose out of the full bent of the tumbler, and thereby releasing it to the action of the main spring.

SIXTH LESSON (FIFTH CONTINUED).

The Bridle.—The bridle secures the tumbler and ear in their proper places in the lock plate; the stud (under the foot of the bridle) fits into the lock plate, and together with the sear and bridle pins keeps the bridle firm and in its place.

The Hammer.—The hammer fits on the squares of the tumbler; it falls with great quickness and force upon the cap, thereby causing the explosion.

The Tumbler.—This is the most important part of the machinery of the lock, it holds the hammer and revolves in the lock plate and bridle; it is made very strong to sustain the force of the main spring; the notches, termed bents, are to retain the hammer in the position of half or whole cock, *i.e.* to "cap" or explode; the half or lower bent in its construction has two angles, one acute, the other obtuse; such a formation is essential, so as to prevent the possibility of the sear nose being released by the action of the trigger; for whatever amount of pressure is applied thereto (if no imperfection exists) would only have the effect of breaking away either the half bent or the nose of the sear.

The full or upper bent is formed so as to allow the nose of the sear to be withdrawn from it with the least possible resistance, consequently these parts must coincide, the edge of the full bent and that of the half bent should be in the same arc of a circle, to ensure the proper action of the tumbler; if the latter projects, the lock will catch at half cock when the sear is released from the full bent; if it is cut away too much, the position of half cock is not secure. The shaft or lever is for the purpose of connecting the main spring to the tumbler by means of the swivel.

The Lock Plate.—This is the foundation of the lock, on which the several parts are fixed.

The fore stud, as before stated, is to secure the catch of the main spring as well as to offer a resistance to ensure its action. The hind stud acts as a break to prevent the tumbler from revolving beyond what has been considered necessary when acted upon by the main spring, whose action is consequently stopped.

To ensure a true and easy action of the lock, the pins, pivots, &c., should be perpendicular to the lock plate.

The bridle pin is flat at the end, the other pins are rounded off. The bridle pin is the longest, the sear next, and the sear spring pin the shortest.

Seventh Lesson (in two parts).
How to guard against the Rifle and Lock getting out of order.
(In two lessons.)

The Pull Off.—It is a great error to suppose that by loosening the sear or any other pins an easier pull off can be ensured; such a measure is only calculated to impair the efficiency of the lock, and increase the wear and tear of the parts which are thereby thrown out of order. A perfect parallelism is most essential to secure the object required.

When the pull off is too great, it proves that the full bent and sear nose are not well adapted to each other, or that the sear spring is too strong.

The alteration that becomes necessary to remedy so serious a defect in a lock must be made by an armourer, and never attempted by the soldier.

Miss Fires.—The non-explosion of the cap is often attributed to the weakness of the main, spring; this may sometimes be the case, but in the majority of instances, the cause may be traced to either dirt at the base of the touch hole, or to the dirty or rusty state of the tumbler axle hole and axle of the tumbler; either of these two causes will impede the fall of the cock, and, as a natural consequence, likewise reduce the force of the blow so necessary to explode the cap.[1] It is therefore of the utmost importance, when cleaning the rifle, to prevent water entering the lock by the axle hole, or dirt remaining in the touch hole.

Wood Bound.—The lock is often said to be wood bound by (what many suppose) the swelling of the wood when the rifle has been exposed in wet weather; such a circumstance, however, too often arises from screwing the lock to the stock by the side nails, beyond what is necessary, and thereby imbedding the parts of the lock into the wood work of the stock.

The side nails (as also the pins of the lock) should not, when screwed home, protrude beyond the outer side of the lock plate. The ends of the nails and pins (except the bridle pin, which is flat), should be rounded off to the lock plate, and present no edges.

---

1. Miss fires are also occasioned by an imperfection in the construction of the communication hole, or the nipple screw being too long; as a consequence, when the nipple is screwed down, part of the communication hole is closed, and thus prevents the powder from getting into the chamber. The armourer alone should remedy such a defect.

EIGHTH LESSON.

The value of the present rifle over the smooth bore musket cannot be too highly estimated; but as its value chiefly depends upon the straightness and perfection of the bore, you must, by bestowing care and attention to all its parts, prove that you regard it as a most valuable and delicate arm, but, if you act otherwise, you will find that the results will not come up to the real powers which the weapon is capable of performing.

The Barrel.

On no account whatever must the rifle be used for carrying any weight, or for any purpose for which it is not legitimately intended; the barrel is very easily bent, and though the injury may be so slight as not to be perceptible to any but a practised viewer, it may be a sufficient cause to destroy the accuracy of the bore, and consequently of its shooting. If you suspect that the barrel is either bent or dented, the circumstance should be reported immediately.

Avoid all useless thumping and rattling of the rifle.

In "Piling Arms," be attentive to the directions given that they may not become unlocked, and fall to the ground.

When on *Guard*, and when "*Guard Turn Out*," is called, be careful not to snatch the rifle from the "arm-rack" too suddenly, the bayonet may become strained, or the stock or barrel injured, particularly at the muzzle, which is about the thinnest part of the barrel, and that part being the point of delivery causes an irreparable damage.

Be very particular when cleaning your rifle not to rub or damage the "*fore-sight*," this you may unthinkingly do, not knowing how greatly such damage may operate against you at the time of trial.

Keep the interior of your barrel perfectly free from rust and other damage, for if rust be allowed to accumulate in the barrel, a resistance will be offered to the passage of the bullet, causing it most probably to "strip," *i. e.* it may pass out of the grooving without attaining the rotatory motion which is actually essential to the accuracy of its flight. Rust in the barrel prevents a true expansion of the bullet, and renders the rifle difficult to load. Rust is caused by the combined effect of moisture and air; the surest way, therefore, of preventing rust in the barrel and to keep the bore perfectly dry, keep the muzzle-stopper in the snap cap on the nipple, and the hammer down, so that the air may be excluded.

Be very careful when skirmishing not to run the muzzle of the barrel into the ground; if you should accidentally do so, immediately ask permission to *fall out*, and remove the dirt; for if you fire the rifle without removing the obstruction, it is liable to burst the barrel in your hand, and do you a very serious injury. If your rifle is not loaded, remove the dirt carefully, and wipe the barrel well out; for if the dirt is rammed down with a charge, it will tear the barrel and destroy the surface of the bore.

Ammunition.

Be very particular that the inside of your pouches is clean, free from dirt or dust, which may adhere to the greased part of the cartridge, and thereby prevent the bullet from going down the barrel with the ease required.

Whenever you have any loose ammunition in your pouch—that is when a packet of ammunition has been broken into—fold up carefully the loose cartridges in paper that they may not shake about and become damaged. For the same reason always pack your ammunition in your pouches in such a manner as to be always ready for service, but having no vacant space between the packets.

Keep your powder dry; this is of the utmost importance, for bear in mind that damp powder will not send a bullet so far as powder that is perfectly dry, and is more liable to cause miss fires. Hence, then, the necessity of keeping the ammunition and percussion caps in a dry and serviceable state.

Whenever you are out on piquet, or other duty, in which your rifle is likely to be exposed to rain when loaded, you may prevent the powder in the barrel from getting wet by removing the percussion cap, and placing the snap cap on the nipple instead. If you have no snap cap, stop up the nipple hole with grease, and let the cock down upon it. If you cannot obtain any grease, drive a peg of soft wood into the nipple, and put the cap on it. Neither the grease nor the peg will in any way impede the action of the cap in igniting the charge, but both will disappear (if the percussion cap has been properly pressed home) in the explosion.

Whenever the grease round the bullet appears to have been removed from the cartridge, the sides of the bullet should be wetted in the mouth before it is put in the barrel, as the saliva will serve the purpose of grease for the time being.

Directions for Dismounting the Rifle.

1st. Put the stopper into the muzzle of the barrel, take out the ramrod, pull the lock up to the half cock, and take it off. To do this hold the rifle horizontally in the left hand, with the lock downwards, unscrew the side nails, and, if the lock should stick, tap lightly on the head of the side nails, so that it may fall into the left hand.

2nd. Place the muzzle on the ground (with the stopper in it), barrel towards the body and upright, partly unscrew the breech nail.

3rd. Reverse the position of the rifle by placing the butt on the ground, the barrel upright and towards the body if the rifle is fitted with spring bands; press upon the springs with the two forefingers of the left hand, and with the forefinger and thumb of the right hand. Slip the bands over the springs, and take them off the barrel. If the rifle is fitted with screw bands, turn the barrel to the front, and unscrew the bands just sufficient as will allow them to be taken easily off the rifle with the left hand.

4th. Reverse the rifle by placing the muzzle on the ground, and take out the breech nail, after which take hold of the muzzle with the right hand, and lift it gently out of the stock; if the barrel will not leave the stock with ease, hold the rifle by the stock in a slanting direction, the barrel downwards, and the left hand at the full extent of the arm, then tap the heel of the butt lightly on the ground, so as to start the tang of the breech pin from the wood.

Directions for Remounting the Rifle Musket.

Put the barrel into its place in the stock, having the breech properly settled; slip on the bands over the springs; touch the screw threads of the breech nail with oil, and screw it into its place (but not quite home). Pull the hammer up to the half bent, and put on the lock, screw home the side nails, and breech nail. Put in the ramrod, and ease the lock.

Directions for Cleaning the Stock.

Rub the stock with a little linseed oil, after which wipe it well, and apply a little bees' wax, more especially round the lock plate, and between the stock and barrel, so as to prevent wet entering either into the lock or between the stock and barrel.

Vide Third Lesson.

Memorandum.

Horse Guards, 6th January, 1856.

The annexed "Directions for Cleaning Rifle Muskets" having been approved by the General Commanding-in-Chief, are promulgated to the Army with a view to the same being generally adopted.

By Command,

(Signed)   G. Wetherall, A.G.

Directions for Cleaning Rifle Muskets,

1st. Place the musket at full cock.

2nd. Pour about a quarter of a pint of clean water into the barrel j in doing this hold the musket in the left hand, in a slanting direction, keeping the muzzle a little below the elbow of the arm with the barrel downwards to prevent any spilt water running between the barrel and the stock.

3rd. Put a piece of rag or tow into the tag and surround it with the same, put it into the barrel *immediately the water is poured in*, and rub it well up and down, forcing the water out of the barrel through the nipple vent, which repeat once or twice.

4th. Wipe the barrel well out with rag or tow until quite clean and dry, and then with an oiled or greased rag.

Note.—By this mode of cleaning, it is expected there will be little liability of the barrels becoming rusted, and seldom any necessity for removing the barrel from its stock, which is always objectionable, even with muskets fitted with the break off.

Wipe out the oil or grease with a clean rag just before firing.

Method of Case Hardening.

Is performed by stratifying the various parts in an iron pan, with animal charcoal, prepared from bone and ivory dust, or old shoes; the whole is then exposed to a full red heat for about an hour, or according to the size of the work; the pan is withdrawn from the fire, and the contents thrown into a bucket of water. The rationale of this operation is that the surface of the iron becomes converted into steel by the absorption of the carbon, and the beautiful colours are produced *by the animal matter remaining in it*, the variegation of the colour being also affected by the quality of the iron.

INSTRUCTIONS TO ASCERTAIN THE STRENGTH OF THE SPRINGS OF A LOCK, AND THE PULL OFF WITH AND WITHOUT TRIGGER.

Weight of main spring, or main spring draws at half cock from 13 to 14 lbs.

Proof—Attach such a weight to the comb of the hammer, when the lock is at half cock, as will just move the hammer or over-balance the main-spring.

Weight of sear spring, or sear spring draws when the lock is at its bearings about 7½ lbs.

When the hammer is down, place a weight to the arm of the sear as will just lift it, or over-balance the sear spring.

Pull of the lock, or pull off without trigger from 13 to 14 lbs.

Place the lock at full cock, then attach a weight to the arm of the sear as will raise the nose out of the full bent to allow the hammer to fall.

Trigger draws, or pull off with trigger from 7 to 8 lbs.

Attach such a weight to the trigger, when the lock is at full cock, as will raise the sear nose out of the full bent to allow the hammer to fall.

The steelyard is used in these experiments or tests.

Browning Arms.

The following ingredients for the browning of arms are to be mixed and dissolved in one gallon of soft water, *viz.*:

6 ozs. spirits of wine.

6 ,, tincture of steel.

2 ,, corrosive sublimate.

2 ,, sweet spirits of nitre.

3 ,, nitric acid.

The mixture is to be kept in glass, not in earthenware bottles.

(*Vide Queen's Regulations*, p. 96, sect. 20.)

Previously to commencing the operation of browning, the barrel is to be made quite bright with emery, or a fine smooth file (but not burnished), after which it is to be carefully cleaned from all greasiness; a small quantity of pounded lime, rubbed well over every part of the barrel, is the best for this purpose; a plug of wood is then to be put into the muzzle of the barrel, the nipple and touch-hole are to be stopped, and the mixture applied to every part with a clean sponge or rag.

After the application of the mixture, the barrel is to be exposed to the air in a warm room for twelve hours, after which time it is to be well rubbed over with a hard hair-brush, or armourer's brush, until the rust is entirely removed. The mixture is then to be applied again in the same manner as before and in six hours the barrel will be sufficiently corroded for the operation of scratch-brushing. The process of scratching off the rust, and applying the mixture, is to be repeated twice or three times a day for four or five days, by which time the barrel will be of a dark brown colour.

The rust which is raised by each successive application of the mixture, is to be always removed at first with the hair-brush, previously to using the scratch-card, as the latter is otherwise found to remove the browning. The operation of scratch-carding is only to commence after the second application of the mixture.

When the barrel is sufficiently browned, and the rust has been carefully removed, it is to be placed in boiling water for three or four minutes, in order that the action of the acid mixture may be destroyed, and the rust prevented from rising again. The barrel while warm is to be rubbed over with sweet oil, or common olive oil. The operation of browning should be conducted in a dry and warm room; a temperature of about seventy degrees is the most favourable.

The operation of browning is to be renewed every two years.

List of Prices to be allowed to Armourer-Serjeants of Regiments for the Repair of Rifled Muskets, Patten. 1853.

DESCRIPTION OF REPAIR.	Arms fitted with solid bands and springs.			Arms fitted with screw bands and rammer springs.		
Stock.	£	s.	d.	£	s.	d.
A new Stock, and re-stocking complete .	0	12	0	0	12	0
Splicing do., including splice . . { long	0	5	0	0	4	0
{ short	0	2	6	0	2	0
A new brass nose cap and fitting . . .	0	0	9	0	0	9
A new brass handle, or guard and fitting .	0	1	2	0	1	2
A new brass heel plate, and do.	0	1	3	0	1	3
A new brass box trigger plate, and re-fitting the trigger	0	1	3	0	1	3
A new brass side nail cap, and fitting . .	0	0	2	0	0	2
A new fore or shaft swivel, with nail and fitting	0	0	5	0	0	5
A do. handle swivel, and fitting . . .	0	0	4	0	0	4
A do. trigger, filed up, fitted, and hardened	0	0	6	0	0	6
A do. stopper for rammer, and fitting . .	0	0	2	0	0	2
A do. side or breech nail, filed, tapped, and hardened	0	0	6	0	0	6
A do. wood screw, fitted, and hardened .	0	0	1	0	0	1
A do. wire pin, and fitting	0	0	0½	0	0	0½
A new rammer-spring, and do. . . .	0	0	0	0	0	9
New bands, fitted and { front with swivel	0	1	5	0	1	5
hardened, or blued { middle	0	0	10	0	0	10
{ back . . .	0	0	10	0	0	10
A new band, spring, and fitting . . .	0	0	4	0	0	0
A do. band screw, with nut and fitting . .	0	0	0	0	0	3
Lock.						
A new lock plate, filed, hardened, and fitted	0	2	6	0	2	6
A do. cock, filed, fitted, and hardened . .	0	2	3	0	2	3
A do. steel sear, filed, fitted, and hardened	0	1	5	0	1	5
A do. bridle, fitted and tempered . . .	0	1	0	0	1	0
A do. main spring, to weigh from 13 to 14 lbs., at half bent, including fitting }	0	2	0	0	2	0

DESCRIPTION OF REPAIR.	Arms fitted with solid bands and springs.			Arms fitted with screw bands and rammer springs.		
	£	s.	d.	£	s.	d.
A do. sear spring and fitting	0	1	0	0	1	0
A do. swivel, fitted and tempered . .	0	0	8	0	0	8
A do. steel nail, filed, tapped, { tumbler	0	0	4	0	0	4
and tempered . . . . { lock of sorts	0	0	3	0	0	3
Barrel.						
A new nipple and fitting	0	·0	6	0	0	6
Clipping the breech pin	0	0	9	0	0	9
A new front sight, filed up and fitted .	0	0	8	0	0	8
A do. elevating sight bed and fittings .	0	2	0	0	2	0
A do. do. do. flap do. do . .	0	2	0	0	2	0
A do. slider for sight do.	0	0	9	0	0	9
A do. cap or top piece for sight and do. .	0	0	8	0	0	8
A do. sight spring and fitting	0	0	8	0	0	8
A do. sight screw and do.	0	0	2	0	0	2
A do. centre pin for joint of sight, and fitting	0	0	2	0	0	2
A do. elevating sight, including soldering on, adjusting, cleaning off, and browning bed of sight . . . . . }	0	7	4	0	7	4
Graduating and making bed of sight . .	0	0	1	0	0	1
Do. do. do. flap do. . .	0	0	2	0	0	2
Browning barrel, including smoothing . .	0	1	0	0	1	0
Bayonet.						
A new bayonet, with locking ring complete, including fitting and adjusting, &c. . . . . }	0	6	6	0	6	6
A new locking ring and fitting . . .	0	1	3	0	1	3
A do. screw or stud for locking ring . .	0	0	2	0	0	2
Setting bayonet when bent	0	0	0	0	0	3
Rammer.						
A new steel rammer complete.	0	2	0	0	2	0
Tapping thread for worm	0	0	1	0	0	1

Vide Queen's Regulations, p. 101.

Directions for Making Cartridges and Gunpowder

MANUFACTURE OF CARTRIDGES.

The following articles for the instruction of soldiers in the making of cartridges is supplied to each barrack by the War Department. *Vide* Instructions of Musketry. One set of implements, *viz.*:

Five tin measures, 2½ drams each.

Five tin funnels (long narrow spouts).

Twelve mandrels of hard wood for cartridge pattern, 1853.

Twelve formers.

One set of tin patterns to show the size and shape of paper for cartridge.

One iron straight edge.

One large knife.

Twenty quires of white paper for inner and outer envelopes.

Six quires of cartridge paper for cylinder of cartridge.

Half bushel of fine sand.

Fifty bullets for rifle musket, 1853.

After cutting the paper according to the patterns A B and C, place the mandrel on the stiff paper, the base even with the side A C; roll the stiff paper on the mandrel as far as B (**Fig. 1**), insert the inner envelope between the roll of the stiff paper, keeping the side A B ¾ of an inch from the base of the mandrel (**Fig. 2**); roll the whole tightly on the mandrel (**Fig. 3**), place it vertically (**Fig. 4**), and fold or twist the remainder of the envelope that overlaps the stiff paper into the hollow in the base of the mandrel (**A**). Make use of the point of the former to close the folds (**Fig. 5**); examine the bottom of the inner case thus formed, to see that there remains no hole for the escape of the powder when charged.

Introduce the point of the bullet well into the aperture of the powder case. Place the mandrel and bullet parallel to the side A B, and the base of the bullet at ½ an inch from the base A C of the outer envelope (**Fig. 6**); press the point of the bullet into the cavity, roll the envelope tightly on the bullet and on the mandrel (**Fig. 7**), twist or fold the remainder of the envelope, and tie it as close as possible to the base of the bullet (**Fig. 8**); place the base of the cartridge on the table, withdraw the mandrel, squeezing the edge of the powder case with the thumb nail of the left hand, and raising up the mandrel with the right hand (**Fig. 9**).

To charge the cartridge, introduce the point of the funnel [1] into the bottom of the case of the cartridge; pour in 2½ drachms of fine grain powder (or sand) from the powder flask; withdraw the funnel, taking care that none of the powder escapes between the case and the envelope, squeeze the top of the cartridge (**Fig. 10**) and twist it inwards, forcing it a little into the case.

When completed, the base of the cartridge (**Fig. 10, A**) must be dipped up to the shoulder of the bullet B in a pot of grease, consisting of one part of tallow to five of bees' wax.

GUNPOWDER

Gunpowder is a composition of sulphur, saltpetre, and pounded charcoal; these three ingredients, when mixed together in proper quantities, form a very inflammable substance. A single spark is sufficient to inflame in an instant the largest mass of this composition.

The expansion, suddenly communicated either to the air lodged in the interstices of the grains of which it consists, or to the nitrous acid which is one of the elements of the saltpetre, produces an effort which nothing can resist; and the most ponderous masses are driven before it with inconceivable velocity.

Sulphur is generally found ready formed in the vicinity of volcanoes, and almost in a state of purity.

Saltpetre, or what is chemically called nitrate of potash, is found in a natural state, but in small quantities. In India it is sometimes found on the surface of the ground, and sometimes on the surface of calcareous walls, the roofs of cellars, and under the arches of bridges. As charcoal is so well known, I shall only observe that the wood selected for the manufacture of gunpowder are the willow, alder, or black dog-

---

1. A copper funnel should always be used when making service cartridges.

wood; the latter is generally used in making the best sporting powder, the two former for Government powder.

Gunpowder is made of the three ingredients already named, which are combined in the following proportions: for each 100 parts of gunpowder, saltpetre 75 parts, charcoal 15, and sulphur 10,

The several ingredients being thus prepared, are sent to the mixing house; there they are ground separately to a fine powder, and carefully weighed in the proper proportions; they are then sifted into a large trough, and well mixed together by the hands. The composition, after being mixed, is taken to the powder mill, which is a light brick or wooden building, having in the centre two stones placed vertically and fixed to an horizontal axis traverse; these mill stones weigh from three to four tons, and make about nine or ten revolutions in a minute; beneath them, and placed in a circular trough, is a smooth cast-iron or stone bed; on this bed the composition is spread out; a sufficient quantity of water is put on it so as to reduce it to a body, but not so much as would form it into a paste. After the mill stones (or runners as they are sometimes called) have made the necessary revolutions over the composition, (which is generally regulated according to the state of the atmosphere,) it is taken off the bed in form of mill cake, and taken to the press house, where it is firmly pressed between plates of copper, worked by means of a powerful screw until it is formed into a hard and firm mass.

After this process it is taken from the press in slabs resembling slate. It is then broken into small pieces with wooden mallets, and taken to the coming house to be granulated; this is done by putting the broken pieces into sieves, the bottoms of which are made of parchment skins or bullocks, skins, prepared like parchment, with round holes punched through them about two-tenths of an inch in diameter. Several of these sieves are secured to a frame. Upon the press cake in each sieve is placed two pieces of *lignum vitae*, six inches in diameter, and two or three inches in thickness; so that when the frame which is attached to the machinery is put into rapid motion, the circular pieces of *lignum vitae* presses upon the powder, breaks the lumps by the velocity in which it moves, and thereby forces it through the holes in grains of several sizes on to the floor. It is then separated from the dust by being sifted through finer sieves, and also to classify the grain.

The grains are next hardened, and the rough edge taken off by friction against each other in barrels, or in a close reel, which is moved in a circular direction, making about forty revolutions in a minute, by this process also the grains become glazed. The object of glazing is to break off the acute angles from the grains, and to give the powder a more finished appearance; it however diminishes rather than increases the strength.

The gunpowder thus corned, dusted, and glazed, is then sent to the stove to be finally dried by an artificial temperature of 140 degrees Fahrenheit, which is suffered gradually to decline. The last process is taking it to the dusting house, where it is sifted into grains of various sizes, and afterwards through a fine wire or canvas reel, to free it from all dust.

If gunpowder is injured by damp in a small degree, it may be recovered by again drying it in a stove; but if the ingredients are decomposed, the nitre must be extracted, and the gunpowder remanufactured.—*Hutton.*

There are several methods of proving and trying the goodness and strength of gunpowder. The following is one by which a tolerably good idea may be formed of its purity, and also some conclusion as to its strength.

Lay two or three small heaps, about a drachm or two of the powder, on separate pieces of clean writing paper; fire one of them by a red hot wire; if the flame ascends rapidly, with a good report, leaving the paper free from specks, and without burning holes in it; and if sparks fly off and set fire to the adjoining heaps, the goodness of the ingredients, and proper manufacture of the powder, may be safely inferred; but if otherwise, it is either badly made, or the ingredients are impure.

Physical causes of its inflammation and exploding.—

Gunpowder being composed of the above ingredients, when a spark falls on this mixture, it sets fire to a certain portion of the charcoal, and the inflamed charcoal causes the nitre with which it is mixed, or in contact, to detonate, and also the sulphur, the combustibility of which is well known. Portions of the charcoal contiguous to the former take fire in like manner, and produce the same effect in regard to the surrounding mass; thus the first portion inflamed, inflames a hundred others; these hundred communicate the inflammation to ten thousand; the ten thousand to a million, and so on. It may be easily conceived that an inflammation, the progress of which is so rapid, cannot fail to extend itself, in the course of a very short time, from the one extremity to the other of the largest mass.—*Hutton*.

The following extract from Mr. W. Greener's *Gunnery, 1858*, may prove interesting, and thereby induce young soldiers to read such works placed at their disposal in the barrack libraries, as will satisfy them of the cause of such a curious fact.

If a train of gunpowder be crossed at right angles by a train of fulminating mercury, laid on a sheet of paper on a table, and the gunpowder lighted by a red-hot wire, the flame will run on until it meets the cross train of fulminating mercury, when the inflammation of the latter will be so instantaneous as to cut off the connection with the continuous train of gunpowder, leaving one-half of the train unignited: (and again),—If the fulminating powder be lighted first, it will go straight on, and pass through the train of gunpowder so rapidly, as not to inflame it at all.'

Mr. Greener says:

....the cause is quite apparent; the rapidity of combustion condenses the air so quickly, as to remove the grains of gunpowder liable to come in contact with the flame, and to form the condensed air into a line of demarcation: for heat cannot be taken up by the air quicker than the atmosphere will convey sounds, and before the heat can evaporate, the explosion is over, and is consequently noiseless.

The history of opinions respecting the explosive force of gunpowder, and all alike pretending to be deduced from experiments, is scarcely less amusing than the hypothesis respecting the cause, although rendered much more marvellous by their extraordinary discrepancy.

John Bernoulli considered the initial force as equal to 100 times the pressure of the atmosphere; whilst Daniel Bernoulli made it 10,000, Bracehus determines it at 450. D'Antone as lying between 1,400 and 1,900, and Ingenhouz at 2,276. According to Dulacy it is 4,000, by Amontons it is estimated at 5,000, and by Lombard it is stated at 9,215. After this there is a rapidly increasing estimate among other experimenters. Monsieur le Genéral de la Martillière representing it at 43,600, Count Rumford at 54,750; and Monsieur Gay de Vernon, who outdoes all his competitors, stating it as making from 30,000 to

80,000.—*Edinburgh Encyclopaedia.*

PERCUSSION POWDERS.

Mr. Blaine, in his *Encyclopaedia of Rural Sports*, says, that the first fulminating powder was composed of oxymuriate of potash, fine charcoal, and sulphur, in various proportions; but a common one as follows. To any given quantity of the first article, add one-eighth of the second, and one-sixteenth of the third.

Dr. Ure, in his *Chemical Dictionary*, gives a very clear condensed account of the chemical properties and method of manufacturing this powder.

A hundred grains of mercury are to be dissolved by heat in an ounce and a half by measure of nitric acid. This solution being poured cold into two ounces (by measure) of alcohol in a glass vessel, heat is to be applied till effervescence is excited. A white vapour undulates on the surface, and a powder is gradually precipitated, which is to be immediately collected on a filter, well washed, and cautiously dried with a very moderate heat.

This powder detonates loudly by gentle heat, or slight friction. Should any attempt be made to make this powder in any considerable quantity, the process must be conducted accordingly.

The fulminating mercury (according to the same authority), should be moistened with thirty *per cent*, of water, then triturated in a mortar, and hereafter mixed with the sixth part of its weight of gunpowder. Matches, caps, &c., thus made to resist damp, and do not corrode the instrument which contains the powder.

They will even take fire after several hours immersion in water, whereas the priming powder, made with chlorate of potash, sulphur, and charcoal, attracts moisture speedily, and corrodes every metal it comes in contact with, consequently is injurious to the gun it is used with.

The Parisian percussion caps are all of them thus made, and are sold at the rate of six *francs* the thousand. *The Bulletin des Sciences Militaires* states, that the cap-match is made thus:—

Having triturated ten parts of fulminating mercury on a marble slab with water by means of a wooden muller, add six parts of gunpowder, and grind the two together.

Of these military cap-matches, each, it is said, contains one grain and a quarter of fulminating mercury, with six-tenths of that weight of cannon powder.

Percussion caps may be used over again if re-primed, by pressing them with a punch to fit, into an iron hole or mould, and a small quantity of percussion powder at the bottom, with a piece of paper pressed over it; or if caps cannot be had, three or four layers of paper cut to size, pressed into the mould with a punch, and primed with powder, similarly to the copper cap, will make a substitute.—*From Instructions on Naval Gunnery.*

A Practical Guide to Squad and Setting-up Drill

S. Bertram Browne

Contents

Preface

Diversity of opinions constantly occurring in reference to the words of command and cautions used throughout the different practices of the Position Drill, (such diversities being caused by instructors not adhering to one uniform system of words of command and caution, whereby the attention of the squad becomes perplexed, and a total departure from the principles and intention of the above drill is the result ;) the author of the following system of instruction has felt the necessity of some defined rules for that purpose.

Where uniformity exists, diverse opinions perish. Hence the author has been induced to arrange this system with a view to establish one uniform mode of carrying out the very important duty of not only training the recruit to become habituated to the erect and unconstrained position of the soldier (previous to his taking arms), but also for the purpose of maintaining an uniformity of practice, that a soldier-like position may be attained and continued.

It may also be made useful to the Volunteers, Police Force, and in Schools and Colleges where military exercises form a part of the studies of those establishments.

For his principal guide the author has consulted several valuable treatises on physical education, and health, written by physicians of eminence; and their unanimous opinions are, that such exercises contained in this small work are of such a nature, and sufficient to keep the body in a perfect state of health.

The author has urged nothing impossible to perform; he has had an opportunity of testing the practicability of the position drill when performed by females, and through judicious management he has had the satisfaction of finding a healthy constitutional improvement, combined with an easy and graceful carriage, and not what the dominion of prejudice would have, that what is intended for the benefit of the one sex must inevitably hurt the other.

It has been the author's principal aim to convey as much information as possible within the limit he has prescribed for himself; and he will have accomplished his views if the following course of instruction should be found useful in promoting the improvement of a system, easily defined, as well for the instructor in carrying out the elementary training of his pupils, and for those pupils in procuring with greater facility instruction for themselves.

The whole of the "Setting-up Drill" is so arranged that schoolmasters and their assistants can readily undertake the drilling of their pupils themselves; and, parents who may object to the drilling of their children at school, will find this little work very useful in enabling them to carry out any portion of the exercises at home, and at hours most convenient and conducive to the health of their children.

It is of the utmost importance, not only to those who are drilled, but to the whole nation, that instructors of sound professional knowledge, and possessing a keen perception of the temper and habits of boys, should be selected to instruct our young male population in military exercises. Superficial qualifications should be ignored, and intrinsic worth be the only standard by which instructors should be weighed, otherwise the good intended by physical education will be nullified, by placing youth under inexperienced and injudicious Instructors.

Introductoty Remarks

The first part of military preparation consists in improving the power of movement and action in the individuals who compose the army; and as the perfection of that power is important to success, it is essential that the principles through which it may be improved be thoroughly understood.

The individual is here considered as a part in a compound instrument, and in order that the parts may correspond in action, and thus act to advantage, it is necessary, not only that they be placed upon a just balance with one another in the primary arrangement, but that they be tried and adjusted relatively in the instrument according to fitness of constitutional power.

The military positions are, or ought to be, attitudes of perfection, according to mechanical rule, so that there may be a facility in concentring and combining exertions for a military purpose. As this is a plain fact, it is important that the young soldier be well set up, according to the military phrase; in common language, placed upon his haunches in such manner that all the joints and joinings of the different parts bear equally and fairly upon each other.

The form of training, or setting up here suggested, is not intended as a mere matter of moulding for the sake of pleasing the eye; it is of positive use; and, in order that the use and end of it may be attained with facility, it is recommended that all young people undergoing a course of military training should have strongly impressed upon their minds the great importance of constantly keeping the body upright when walking, standing, or sitting down

To the haunches, as to the common centre of motion of the human figure, are ultimately referred all the movements performed in military tactics. As just poise is important to the correct execution of action, whatever it may be, it is necessary that poise or balance be studied, understood, and tried in all positions. It is clear that bodily action cannot possess compass, power and ease, unless the movement be made justly and correctly upon the haunches, as on a central pivot. If the movement have not compass, power and ease, force and endurance will not be found in the military act.

The human figure is erect, when man attains a certain point of growth, and assumes locomotion, it is maintained erect, or it is moved from its erect posture by the action of the muscles. The nearer the figure is to a perpendicular—that is, the more equally the various pieces of which the vertebral column consists bear upon each other, the more easily will the balance be preserved under movement. The erect position is maintained by the action of the muscles, which, as they act in succession, relieve each other, and, in consequence of such relief, the action, though often repeated, is sustained with comparatively little fatigue.

Too many methods cannot be used to improve the carriage of the soldier. In all the positions which tend to give freedom to the muscles, particular attention should be paid that the body is not thrown backwards instead of forward; for should the body be allowed to be thrown back, it will counteract every true principle of movement, and cause unsteadiness in a large body.

The instructors to whom the duty of drill is intrusted should possess an accurate knowledge of the part each has to teach, and evince such a clear, firm, and concise manner of conveying their instructions as will command from their pupils a perfect attention to their directions.

Preparatory to commencing the "Extension Motions," open order will be taken by each recruit stretching out his right arm, and keeping that distance from his right-hand man.[1]

Throughout the "Extension Motions," the word "*Steady*" is not to be given as a word of command: it must only be used as a caution; *viz.*, to prevent a continuation of the motions when being performed without any separate word of command.

When the squads are to resume the position of "Attention," a successive number must be given instead of the word "*Steady*."

The word "*Steady*" may also be given at the completion of any motion in a practice, for the purpose of correcting faults; but, in doing so, the Instructors must be very quick, so as to avoid confusion.

Steadiness should be strictly enforced (on a motion being stopped,)

1. *Vide*—"Formation of the Squad for Drill;" and Section 3, "Squad Drill."

and particular attention paid by instructors that every man stands in the correct position of the last motion.

At all times the caution "*Steady*," should be given in a quick, decisive manner.

At the commencement of each practice, the word of command should be given firm and quick.

Each practice should be distinctly stated, and one pause of the slow time of march allowed between each motion, so that every man may feel the motion complete.

"Stand at Ease" may be given occasionally.

It is not necessary or conducive to the strengthening of the muscles to remain too long in one position; therefore a frequent practice of motions is more essential than a long continuation of them, as the muscles in action soon become fatigued, and require to be relieved by varying the motions, so as to bring other muscles into play.

The squad is formed preparatory to commencing the "Extension Motions" by the instructors advancing the front rank (*i.e.* if they are two deep) four paces, and then bring the right files two paces to the front, the word of command "*Half-Right Turn*" should then be given, the squad waiting for the word before they turn. The instructors should then be very particular that each man makes his half turn correctly, so that each individual's shoulder may conform correctly to the right file. If this conformation be strictly attended to, the shoulders of each file will be parallel from right to left; it will also prevent a distortion of the body, which is generally the case when the squads are performing the motions without separate words of command, having then to take the time from the right file, and if they are not placed at first in a proper position, they are compelled, in watching the right files, to turn the body, thereby losing the correct position of a soldier.

The "Extension Motions" in slow and quick time may be performed with and without any separate words of command; but they should always be gone through the first time with separate words of command. Without separate words of command the instructors will have a better opportunity of detecting any careless or lazy man; also to observe that all the motions lead correctly into each other, that they are performed with correctness, and that no motion is slurred over.

There is one particular object which is absolutely necessary should have the first attention of instructors; *viz.*, if the day on which they are performing the extension motions should be windy, the instructor should invariably face his squad nearly opposite the point from which the wind blows, and then take his position, with his back directly opposite the point. By this method the words of command will be distinctly heard, and the men be more steady than if they were faced otherwise.

Particular attention must be paid in dressing, that the body is kept perfectly steady (in the several turnings of the eyes) and square to the front.

In schools the whole of the back stick, and the club exercise, will be found beneficial, as it calls into play the muscles of the chest, trunk, and arms.

All exercise should always be proportioned in extent to the constitution. The length of a lesson ought never to be measured by the hour-hand of a clock, instead of its effect upon the constitution.

Instructors should carefully avoid great fatigue, and always adopt the kind, degree, and duration of each particular exercise; and to remember, that the points at which these results are to be obtained, is not the same in any two individuals, and can be derived only by experience and careful observation.

All indirect exercises that have a tendency to expand the chest, and call the organs of respiration into play, ought especially to be preferred, as they excite the lungs themselves to freer and fuller expansion.

When either from hereditary predisposition, or accidental causes, the chest is unusually weak, every effort should be made to form the growth and strength of the lungs by the habitual use of such of the following exercises as can be more readily practised: the earlier they are resorted to and the more steadily they are pursued, the more certainly their beneficial results will be experienced.

Although the club exercise is beneficial, care must be taken not to counteract such benefit, by the weight of the club being disproportioned to the weak frames that use them; in which case, they pull down the shoulders by dint of mere dragging

When this, or any other exercise, is resorted to in the house, the windows ought to be thrown open, so as to make the nearest approach to the external air.

Short and frequent drills are preferable to long lessons, which exhaust the attention both of the instructor and pupil.

The weight of clubs for the use of soldiers being five, six, seven, eight, and nine pounds; those used in schools may weigh from three to five pounds.

Recruit or Squad Drill

General Rules

1

Instruction of the Recruit.—

1. The instructors must be clear, firm, and concise in giving their directions. They must allow for the different capacities of the recruits, and be patient where endeavour and goodwill are apparent.

2. Recruits should fully comprehend one part of their drill before they proceed to another. When first taught their positions, they should be properly placed by the instructor; when more advanced, they should not be touched, but taught to correct themselves when admonished. They should not be kept too long at any one part of their exercise.

2

Duration of Drills, &c. Short and frequent drills are preferable to long lessons, which exhaust the attention both of the instructor and recruit. The recruits should be moved on progressively from squad to squad according to their merit, so that the quick, intelligent soldier may not be kept back by men of inferior capacity. To arrive at the first squad, should be made an object of ambition to the young soldier.

3

*Mutual Instruction.—*A system of mutual instruction will be practised amongst recruits; it gives the young soldier additional interest in his drill, and prepares him for the duties of a non-commissioned officer. Recruits should, in turn, be called out to put their squad through the exercises which have been practised, and encouraged to correct any error they may observe lists of those who show talent for imparting instruction should be kept, for reference, by the captains, and in the orderly room.

4

Words of Command.—

1. Every command must be loud, and distinctly pronounced, so as to be heard by all concerned.

2. Every command that consists of one word must be preceded by a caution: the caution, or cautionary part of a command, must be given slowly and distinctly; the last or executive part, which, in general, should consist of only one word or sylla-ble, must be given sharply and quickly; as *Company—Halt: Half Right—Turn.* A pause of slow time will invariably be made be-tween the caution, or cautionary part of a command, and the executive word.

3. The words given in the *Extension Motions* and balance step (**Ss. 5, 11**) must be given sharply, or slowly and smoothly, as the nature of the motion may require.

4 When the last word of a caution is the signal for any *prepara-tory* movement, it will be given as an executive word, and sepa-rated from the rest of the command by a pause of slow time; thus, *Right—Form, Quick—March,* as though there were two separate commands, each with its caution and executive word.

5. When the men are in motion, executive words must be com-pleted as they are commencing the pace which will bring them to the spot on which the command has to be executed. The cautionary part of the word must, therefore, be commenced accordingly.

6. Officers, and non-commissioned officers, should frequently be practised in giving words of command. It will be found a good plan to practise several officers, or non-commissioned of-ficers, together in giving words of command, first in succession, then simultaneously; the time and pitch being first given by the instructor.

RECRUIT OF SQUAD DRILL, WITH INTERVALS.

1. A few men will be placed in line (that is, side by side) at arm's length apart; while so formed, they will be termed a "Squad with Intervals."

2. If necessary, the squad may consist of two such lines of men; in

which case the men in the second line will cover the intervals between the men in the first, so that in marching they may take their own points, as directed in **S. 10.**

3. Recruits should in the first instance be placed by the instructor without any dressing: when they have learned to dress, as directed in **S. 3,** they should be taught to fall in as above described, and then to dress and to correct their intervals: after they have been instructed as far as **S. 22,** they may fall in in single rank, and then, if required to drill with intervals, be moved as directed in **S. 24.**

Recruits will, if possible, be instructed singly as far as **S. 22.**

S. 1. POSITION OF THE SOLDIER.

The exact squareness of the shoulders and body to the front is the first and great principle of the position of a soldier. The heels must be in line and closed; the knees straight; the toes turned out, so that the feet may form an angle of 45 degrees; the arms hanging easily from the shoulder, the hand open, thumb to the front and close to the forefinger, fingers lightly touching the thigh: the hips rather drawn back, and the breast advanced, but without constraint; the body straight and inclining forward, so that the weight of it may bear principally on the fore part of the feet; the head erect, but not thrown back, the chin slightly drawn in, and the eyes looking straight to the front.

When the soldier falls in for instruction, he will be taught to place himself in the position above described.

N.B.—The words in the margin printed in italics, are the commands to be given by the instructor.

S. 2. STANDING AT EASE.

Soldiers will first be taught the motions of standing at ease by numbers, then judging the time.

1. By Numbers.
Caution,—Stand at Ease, by Numbers.

One.—On the word "*One*," raise the arms from the elbows, left hand in front of the centre of the body, as high as the waist, palm upwards; the right hand as high as the right breast, palm to the left front; both thumbs separated from the fingers, and the elbows close to the sides.

Two.—On the word "*Two*," strike the palm of the right hand on that of the left, drop the arms to their full extent, keeping the hands together, and passing the right hand over the back of the left as they fall; at the same time draw back the right foot six inches, and slightly bend the left knee.

When the motions are completed, the arms must hang loosely and easily, the fingers pointing towards the ground, the right thumb lightly held between the thumb and palm of the left hand; the body must incline forward, the weight being on the right leg, and the whole attitude without constraint.

Squad-Attention.—On the word "*Attention*," spring up to the position described in **S. 1.**

2. Judging the Time.
Caution.—Stand at Ease, judging the Time.

Stand-at-Ease.—On the word "*Ease*," go through the motions described in the standing at ease by numbers, distinctly but smartly, and without any pause between them.

Squad-Attention.—As before.

If the command "Stand at Ease" is followed by the word "Stand Easy," the men will be permitted to move their limbs, but without quitting their ground, so that on coming to "Attention" no one shall have materially lost his dressing in line. If men are required to keep their dressing accurately, they should be cautioned not to move their left feet.

On the word "Squad" being given to men standing easy, every soldier will at once assume the position of standing at ease.

S. 3. DRESSING A SQUAD WITH INTERVALS.

Eyes-Right.—On the words "*Eyes-Right*" the eyes will be directed to the right, the head being slightly turned in that direction.

Dress.—On the word "*Dress*," each soldier, except the right-hand man, will extend his right arm, palm of the hand upwards, nails touching the shoulder of the man on his right; at the same time he will take up his dressing in line by moving, with short quick steps, till he is just able to distinguish the lower part of the face of the second man beyond him; care must be taken that he carries his body backward or forward with the feet, keeping his shoulders perfectly square in their original position.

Eyes-Front.—On the words "*Eyes-Front*," the head and eyes will be turned to the front, the arm dropped, and the position of the soldier, as described in **S. 1**, resumed.

Dressing by the left will be practised in like manner.

S. 4. TURNINGS.

In going through the turnings, the left heel must never quit the ground; but the soldier must turn on it as on a pivot, the right foot being drawn back to turn the body to the right, and carried forward to turn it to the left; the body must incline forward, the knees being kept straight.

In the first of all the following motions, the foot is to be carried back, or brought forward, without a jerk, the movement being from the hip; so that the body may be kept perfectly steady until it commences to turn.

Right-Turn.—On the word "*Turn*," place the hollow of the right foot smartly against the left heel, keeping the shoulders square to the front.

Two.—On the word "*Two*" raise the toes, and turn a quarter circle to the right on both heels, which must be pressed together.

Left-Turn.—On the word "*Turn*," place the right heel against the hollow of the left foot, keeping the shoulders square to the front.

Two.—On the word "*Two*," raise the toes, and turn a quarter circle to the left on both heels, which must be pressed together.

Right about-Turn.—On the word "*Turn*," place the ball of the right toe against the left heel, keeping the shoulders square to the front.

Two.—On the word "*Two*" raise the toes, and turn to the right about on both heels.

Three.—On the word "*Three*," bring the right foot smartly back in a line with the left.

Left about-Turn.—On the word "*Turn*," place the right heel against the ball of the left toe, keeping the shoulders square to the front.

Two.—On the word "*Two*," raise the toes, and turn to the left about on both heels.

Three.—On the word "*Three*," bring up the right foot smartly in a line with the left.

Half-Right (or Left)-Turn.—On the word "*Turn*," draw back (or advance) the right foot one inch.

Two.—On the word "*Two*," raise the toes and turn half right (or left) on both heels.

Three-quarters-Right (or Left) about-Turn. Two. Three.—Make a three-quarters turn in the given direction in the same manner as in turning about.

Squad-Front.—After any of the foregoing turnings, the word "Front" may be given, on which the whole will turn, as accurately as possible, to their former front.

When the soldier has previously turned about, he will always front by the right about. But if he has turned to the three-quarters right about, he will front by the three-quarters left about; and *vice versa*.

At squad drill with intervals, the turnings will always be done by numbers, except when the word "Front" is given, in which case the soldier will judge the time, which must be a pause of slow time after each motion.

S. 5. Extension Motions.

In order to open his chest, and give freedom to his muscles, the soldier will be practised in the following extension motions.

Men formed in squads with intervals will be turned a half turn to the right, before commencing these practices.

Caution,—First Practice, (**Plate 11**.)

One.—On the word "*One*," bring the hands, at the full extent of the arms, to the front, close to the body, knuckles downwards, till the fingers meet at the points; then raise them in a circular direction over the head, the ends of the fingers still touching and pointing downwards so as to touch the forage cap, thumbs pointing to the rear, elbows pressed back, shoulders kept down. (Figs. A and B.)

Two.—On the word "*Two*," throw the hands up, extending the arms smartly upwards, palms of the hands inwards; then force them obliquely back, and gradually let them fall to the position of "Attention," elevating the neck and chest as much as possible. (Fig. C.)

Three.—On the word "*Three*," raise the arms outwards from the sides without bending the elbow, pressing the shoulders back, until the hands meet above the head, palms to the front, fingers pointing upwards, thumbs locked, left thumb in front. (Fig. D.)

Four.—On the word "*Four*" bend over until the hands touch the feet, keeping the arms and knees straight; after a slight pause, raise the body gradually, bring the arms to the sides, and resume the position of "Attention." (Fig E.)

N.B.—The foregoing motions are to be done slowly, so that the muscles may be exerted throughout.

Caution,—Second Practice, (Plate 12)

One.—On the word "*One*" raise the hands in front of the body, at the full extent of the arms, and in line with the mouth, palms meeting but without noise, thumbs close to the forefingers. (Fig. A.)

Two.—On the word "*Two*," separate the hands smartly, throwing them well back, slanting downwards; at the same time raise the body on the fore part of the feet. (Fig. B.)

One.—On the word "*One*" bring the arms forward to the position above described, and so on.

Three—On the word "*Three*," smartly resume the position of "Attention."

In this practice, the second motion may be continued without repeating the words "One," "Two," by giving the order "Continue the Motion;" the squad will then take the time from the right-hand man: on the word "Steady," the men will remain at the second position, and on the word "Three" they will resume the position of "Attention."

Caution,—Third Practice. (Plate 13.)

The squad will make a second half turn to the right before commencing the third practice.

One.—On the word "*One*," raise the hands, with the fists clenched, in front of the body, at the full extent of the arms, and in line with the mouth, thumbs upwards, fingers touching. (Fig. A.)

Two.—On the word "*Two*," separate the hands smartly, throwing the arms back in line with the shoulders, back of the hand downwards.

Three.—On the word "*Three*," swing the arms round as quickly as possible from front to rear.

Steady.—On the word "*Steady*" resume the second position.

Four.—On the word "*Four*," let the arms fall smartly to the position of "Attention."

This practice should also be performed with clubs.

S. 6. SALUTING,

Soldiers will be practised in saluting, first by numbers, then judging the time; being turned to the right for the right-hand salute, to the left for the left-hand salute.

Caution,—Right-hand Salute, by Numbers.

One.—On the word "*One*," bring the right hand smartly, but with a circular motion, to the head, palm to the front, point of the forefinger one inch above the right eye, thumb close to the forefinger; elbow in line, and nearly square, with the shoulder; at the same time, slightly turn the head to the left.

Two—On the word "*Two*" let the arm fall to the side, and turn the head to the front.

Caution,—Right-hand Salute, judging the Time,

Right-hand Salute.—On the word "*Salute*" go through the two motions described in "*One*" and "*Two*."

Soldiers will be taught to salute with the left hand in like manner.

Soldiers, if standing still when an officer passes, will turn towards him, come to "Attention," and salute; if sitting, they will rise, stand at "Attention," and salute. When a soldier addresses an officer, he will salute, and halt two paces from him. When walking, soldiers will salute an officer as they pass him, commencing their salute four paces before they come up to him; they should therefore be practised in marching, two or three together, round the drill-ground, saluting points placed on either side of them, care being taken that they always salute with the hand furthest from the point saluted: when several men are together, the man nearest to that point will give the time.

Soldiers will invariably salute anybody they know to be an officer, whether he is in uniform or not.

MARCHING.

S. 7. LENGTH OF PACE.

In slow or quick time the length of a pace is 30 inches, except in "stepping out," when it is 33 inches, and in "stepping short," when it is 21.

In "double time" the length of a pace is 33 inches.

The length of the side step is 12 inches.

N.B.—When a soldier takes a side pace to clear or cover another, as in forming four deep, which will be hereafter described, the pace will be 24 inches.

S. 8. Cadence.

In slow time, 75 paces are taken in a minute. In quick time, 116 paces, making 96 yards 2 feet in a minute, and 3 miles 520 yards in an hour. In double time, 165 paces, making 151 yards 9 inches in a minute, and 5 miles 275 yards in an hour.

The length of the pace in marching will be corrected with the pace stick, the accuracy of which should occasionally be tested by measurement.

S. 10[1]. Position in Marching.

In marching, the soldier must maintain the position of the head and body as directed in **S. 1**. He must be well balanced on his limbs. His arms and hands must be kept steady by his sides; care being taken that the hand does not partake of the movement of the leg. The movement of the leg must spring from the haunch, and be free and natural.

Both knees must be kept straight, except while the leg is being carried from the rear to the front, when the knee must necessarily be a little bent, to enable the foot to clear the ground. The foot must be carried straight to the front, and, without being drawn back, placed softly on the ground, so as not to jerk or shake the body; the toes turned out at the same angle as when halted.

Although several men may be drilled together in a squad with intervals, they must act independently and precisely as if they were being instructed singly. Each soldier must be taught to march in a straight line, and to make correct pace, both as regards length and cadence, without reference to the other men of the squad.

Before the squad is put in motion the instructor will take care that the men are square individually and in correct line with each other. Each soldier must be taught to take up a straight line to his front by first looking down the centre of his body between his feet, then fixing his eyes upon some object on the ground straight to his front at a distance of about 100 yards; he will then observe some nearer point in the same straight line, such as a stone, tuft of grass, or other casual object about 50 yards distant.

S. 11. Balance Step.

The object of the balance step is to teach the soldier the free movement of his legs, preserving at the same time perfect squareness of shoulders, and steadiness of body; no labour most be spared to attain

---

1. N. B. There is no S. 9 in original book.

this object, which forms the very foundation of correct marching. The Instructor must be careful that the soldier keeps his body well forward, and his shoulders perfectly square, during these motions.

1. Without advancing.

Caution,—Balance Step, commencing with the Left Foot.

Front.—On the word "*Front*," the left foot will be raised from the ground by a slight bend of the knee, and carried gently to the front, without a jerk, the knee being gradually straightened as the foot is carried forward; the foot to be turned out at the same angle as when halted, the sole parallel to and clear of the ground, the heel just in advance of the line of the right toe.

Rear.—On the word "*Rear*," given when the body is steady, the left foot will be brought gently back without a jerk, till the toe is in line with the right heel, clear of the ground; the left knee to be a little bent.

Front. Rear.—When steady, the words "*Front*" and "*Rear*" will be repeated several times, and the motions performed as above described.

Halt.—On the word "*Halt*," which should always be given when the moving foot is in rear, that foot will be brought to the ground in a line with the other.

The Instructor will afterwards make the soldier balance upon the left foot, carrying the right foot forward and backward.

Standing on one leg and swinging the other backward and forward without constraint, is an excellent practice.

2. Advancing.

Caution,—Balance Step, advancing on the word "Forward."

Front.—On the word "*Front*," the left foot will be carried to the front, as described in No. 1.

Forward.—As soon as the men are steady in the above position, the word "*Forward*" will be given, on which the left foot will be brought to the ground at 30 inches distance from heel to heel, toes turned out at the same angle as when halted; and the right foot will immediately be raised and held extended to the rear, toe in line with the left heel, the right knee to be slightly bent. Great care must be taken that the toes remain throughout at the proper angle; that the body accompanies the leg, and that the inside of the heel is placed on the imaginary straight line that passes through the points on which the soldier is marching; that the body remains straight, but inclining forward; that the head is erect, and turned neither to the right nor left.

Front.—On the word *"Front,"* the right foot will be brought forward; and so on, alternately.

Halt—On the word *"Halt"* which should always be given when the moving foot is to the front, that foot will complete its pace, and the rear foot will be brought up in line with it.

S. 12. THE SLOW-MARCH.

The three most important objects in this part of the drill are cadence, length of pace, and direction.

Slow-March.—The time having been given on the drum, on the word *"March,"* the left foot will be carried 30 inches to the front, as directed in **S. 10**; the right foot will then be carried forward in like manner, and so on, alternately.

The soldier must be thoroughly instructed in this step, as an essential preparation for arriving at accuracy in the paces of greater celerity.

S. 13. THE HALT.

Squad-Halt.—On the word *"Halt,"* the moving foot will complete its pace, and the rear foot be brought up in line with it.

It is a general rule that after the word "Halt," the men, whatever their position, will stand perfectly steady, unless ordered to "Dress."

S. 14. STEPPING OUT.

Step-Out.—When marching in slow time, on the words *"Step-Out,"* the soldier will lengthen his pace to 33 inches by leaning forward a little, but without altering the cadence.

This step is used when a slight increase of speed, without an alteration of cadence, is required; on the words "Slow"—"Step," the pace of 30 inches will be resumed.

S. 15. STEPPING SHORT.

Step-Short.—On the words *"Step-Short,"* the foot advancing will finish its pace, and afterwards each soldier will take paces of 21 inches until the word *"Forward"* is given, when the usual pace of 30 inches will be resumed.

This step is used when a slight check is required.

S. 16. MARKING TIME.

Mark-Time—On the words "*Mark-Time*," the foot then advancing will complete its pace, after which the cadence will be continued, without advancing, by raising each foot alternately about three inches from the ground, keeping the body steady; on the word "*Forward*," the usual pace of 30 inches will be resumed.

From the halt, the word of command will be "Slow." "Mark"—"Time."

S. 17. STEPPING BACK.

Step-Back Slow-March. Halt.—In stepping back, the pace will be 30 inches. Soldiers must be taught to move straight to the rear, preserving their shoulders square to the front and their bodies erect. On the word "Halt," the foot in front will be brought back square with the other.

A few paces only of the step back can be necessary at a time.

S. 18. CHANGING FEET.

Change-Feet.—To change feet in marching, the advancing foot will complete its pace, and the ball of the rear foot will be brought up quickly to the heel of the advanced one, which will instantly make another step forward, so that the cadence will not be lost; in fact, two successive steps will be taken with the same foot.

This may be required when any part of a battalion, or a single soldier, is stepping with a different foot from the rest.

S. 19. THE QUICK-MARCH.

The cadence of the slow march having become perfectly familiar to the soldier, he will be taught to march in quick time.

Quick-March.—On the word "*March*," the Squad will step off together, with the left foot, observing the rules given in **S. 10.**

When a soldier is perfectly grounded in marching in quick time, all the alterations of step, and the marking time, and changing feet, laid down for the slow march, will be practised in quick time.

S. 20. THE DOUBLE-MARCH.

Double-March.—The time having been given on the drum, on the word "*March*," the men will step off together with the left foot; at the same time raising their hands as high as the waist, carrying back the elbows and clenching the fists, the flat part of the arm to the side;

the head to be kept erect, and the shoulders square to the front: the knees being more bent, and the body more advanced, than in the other marches. The instructor will be careful to habituate the soldier to the pace of 33 inches.

Squad-Halt.—As in **S. 13**, at the same time dropping the hands and extending the fingers.

The soldier will be taught to mark time in the double cadence in the same manner as in the slow and quick.

S. 21. THE SIDE OR CLOSING STEP,

Soldiers will first be taught the side step by numbers, then judging the time.

1. By Numbers.
Caution,—Right Close, by Numbers.

One.—On the word "*One,*" the right foot will be carried 12 inches to the right, the shoulders and face being kept perfectly square to the front, and the knees straight.

Two.—On the word "Two," the left foot will be closed smartly to the right foot, heels touching.

One—The word "*One*" being repeated, the right foot will be carried on 12 inches as before described, and so on.

Squad-Halt.—When the word "*Halt*" is given, the left foot will be closed to the right, as on the word "*Two.*"

2. Judging the Time.
Caution,—Right Close, judging the Time,

Right Close, Quick-March.—On the word "March" each man will carry his right foot 12 inches direct to the right, and instantly close his left foot to it, thus completing the pace; he will proceed to take the next pace in the same manner: shoulders to be kept square, knees not bent, unless on rough or broken ground. The direction must be kept in a straight line to the flank, neither inclining to the front nor rear.

Squad-Halt.—On the word "*Halt*" the men will complete the pace they are taking, and remain steady.

Soldiers will be practised in closing to the left by numbers, and judging the time, in like manner.

Soldiers will also be practised in taking any given number of paces to either flank, and then halting without word of command; the command to be given thus: "Three paces Right Close, Quick-March."

S. 22. Turning when on the March,

Soldiers will be practised in turning to the right or left, in making a half-turn to the right or left, and in turning to the right or left about, on the march.

Right-Turn.—

1 *Turning to the Right, and then to the Front.*—On the word "*Turn,*" which should be given as the left foot is coming to the ground, each soldier will turn in the named direction, and move on at once, without breaking his pace.

Front-Turn.—On the word "*Turn*" which should be given as the right, foot is coming to the ground, each soldier will turn again to the front, and move on without checking his pace.

Left-Turn. Front-Turn.—

2. *Turning to the Left, and then to the Front.*—Soldiers will turn to the left in like manner, the word "Turn" being given as the right foot is coming to the ground; after which they will turn to the front, the word "Turn" being given as the left foot is coming to the ground.

A soldier will always turn to the right on the left foot; and to the left on the right foot. If the word "Turn" is not given as the proper foot is coming to the ground, the soldier will move on one pace more and then turn.

3. *Making a Half Turn to the Right, or Left.*—Soldiers will also be practised in making a half turn to the right or left, and then moving on (without checking their pace) in a diagonal direction, taking up fresh points, at once, to march on.

4. *Turning to the Right, or Left about.*—Soldiers will also be taught to turn about on the march, which must be done by each man on his own ground, in three paces, without losing the cadence. Having completed the turnabout, the soldier will at once move forward, the fourth pace being a full pace as before.

Squad Drill, in Single Rank.

S. 23. Directing and Reverse Flanks.

As explained in the Definitions.

S. 24. Formation of the Squad in Single Rank.

At this stage of the drill, a few soldiers will be formed in single rank without intervals, that is, nearly touching each other. Each man is allowed a space of 24 inches.

The right-hand or left-hand man being first placed, the remainder will fall in in line one after the other, closing lightly towards him, turning the elbow slightly outwards. Soldiers must be carefully instructed in "The Touch," as, in this formation, it is the principal guide when marching. Each man when properly in line, should be able to feel his right or left hand man at the elbow; the body must be preserved in the position described in **S. 1**.

When a squad in single rank is required to drill with intervals, the instructor will direct the odd numbers to take one pace forward, the even numbers to step back one pace.

S. 25. DRESSING WHEN HALTED.

Soldiers will first be taught to dress man by man, then together.

In dressing, each soldier will glance towards the flank to which he is ordered to dress, with a slight turn of the head, as directed in **S. 3**: he must carry his body backward or forward with the feet, moving to his dressing with short quick steps; bending backward or forward must be avoided; his shoulders must be kept perfectly square, and the position of the soldier, as described in the preceding section, retained throughout.

Two men on the right and one on the left, a pace and a half to the Front. Slow-March.—

1. *Dressing Man by Man.*—Preparatory to teaching a squad to dress by the right, the instructor will order the two men on the right, and one on the left, to take a pace and a half to the front; having completed his pace and a half, the right-hand man will take four side paces to his right; and the three points thus placed will raise their right arms from the elbow, at right angles to their bodies.

Man by Man, by the Right—Dress up.—The instructor, having ascertained that the points are in line, will order his squad to dress up man by man. The third man from the right will take one pace to his front with the left foot, and shuffle up into line in the manner already described: as soon as he is steady, the next man will proceed in like manner, and so on to the left. The faces of the men, not their breasts or feet, are the line of dressing. Each man is to be able just to distinguish the lower part of the face of the second man beyond him.

Eyes-Front.—When the instructor is satisfied that the line is correct, he will give the words "*Eyes-Front,*" on which the men will turn their heads and eyes to the front, the three points will drop their hands, and the right-hand man will close on the squad.

In like manner the squad must be taught to dress up, man by man, by the left; also to dress back, man by man, by the right and left.

2. *Dressing together,*—The men must next be taught to dress forward and backward, taking the pace together, but shuffling up or back in succession, the same points being given as in dressing man by man. The words of command will be "Squad, By the Right" (or "Left")— "Dress Up," or "By the Right" (or "Left")—"Dress Back.".

3. *Dressing without Points.*—When soldiers are on the alignment they have to occupy, and their dressing is simply to be corrected, the words "Right" (or "Left")—"Dress," or after the word "Halt," the word "Dress" only, will be given, on which they will shuffle up or back to their places successively, commencing with the man on the flank from which they are dressed.

When no man is placed for that purpose, the instructor should invariably fix upon some casual object on which to dress his line.

It will be found most useful to accustom men to dress on an alignment and parallel or perpendicular, but oblique, to any well-defined adjacent line, such as the side of a square parade ground.

S. 26. TURNINGS,

The soldier will next practise in single rank, judging the time, the turnings he has been taught by numbers. Men are never unnecessarily to stand turned to the rear.

S. 27. MARCHING TO THE FRONT AND REAR.

The soldier will next practise in single rank the different marches and varieties of step which he has learned singly, or in squad with intervals; the same general rules being observed.

Before a squad is ordered to march, the directing flank must be indicated by the caution, "By the Right," or "By the Left."

During the march, care must be taken that neither the head nor the eyes are ever turned towards either flank; that the dressing is kept by the touch; and that the shoulders are kept perfectly square, and the body steady.

The squad will first be taught to march straight to the front, both by the right and left, in slow and quick time; it will then be practised in all the varieties of step, and in marking time, in both cadences; after which it will be exercised in the double time.

The soldier will be practised in changing the pace, without halting, from slow to quick, and from quick to slow time; also from quick to double, and from double to quick: in the case last mentioned, on the word "Quick," the arms will be dropped and the fingers extended.

The instructor should occasionally remain halted in rear of the man on the directing flank; and, by fixing his eyes on some distant object, ascertain if the squad is marching straight to its front.

When a soldier finds himself a little behind, or before, the other men of his squad, he must be taught to recover his place in the rank gradually, and not to jump or rush to it, which would make him unsteady and spoil the marching of the rest of the squad.

S. 28. A Single Rank at the Halt, changing Front.

Right-Wheel. Slow-March.—

1. *By Wheeling.*—On the word "*March*," the right-hand man, called the pivot man, will mark time, turning gradually with the squad, to the new front; the remainder will step off, the whole turning their eyes to the left (the wheeling flank), except the left-hand man, who will look inwards, and step the usual pace of 30 inches, the other men regulating their length of pace according to their distance from the pivot flank. During the wheel, each man must touch lightly, as explained in **S. 27,** towards the pivot flank, keeping his shoulders square in line; crowding must be carefully avoided; each man must yield to any pressure that may come from the pivot flank, and resist all pressure coming from the outer flank.

Squad-Halt. Dress.—On the word "*Halt*," which may be given at any period of the wheel, the men will halt and turn their eyes to the front; on the word "*Dress*," they will take up their dressing by the right, as described in **S. 25, No. 3.**

Eyes-Front.—On the word "*Front*," the men will turn their heads and eyes to the front.

A squad will wheel to the left in like manner.

Nothing will sooner tend to enable the recruit to acquire the length of step proportioned to his distance from the pivot, than continuing the wheel without halting for several revolutions of the circle.

When men are required to wheel to the rear of the alignment they occupy, they will be turned about, and then wheeled as above directed, receiving the words "Halt," "Front-Dress," followed by "Eyes-Front," when in position.

After wheeling has been taught in slow time, it will be practised in quick and double time.

Right-Form. Quick-March.—

2. *By File formation*.—On the word "*Form*," the right-hand man will turn to the right; the remainder will make a half-turn to the right. On the word "*March*," all, except the right-hand man, will step off; each man, glancing to the right, will move, at the usual length of pace, by the shortest line, to his place in the new front, and take up his dressing by the right.

Eyes-Front,—On the word "*Front*," the men will turn their heads and eyes to the front.

A squad will form to the left in like manner.

This formation will be practised in double as well as in quick time.

S. 29. A Single Rank, on the March, changing Direction.

Right-Wheel,—On the word "*Wheel*" the men will wheel to the right, as explained in **S. 28**, No. 1; the pivot man turning gradually with the Squad.

Forward.—On the word "*Forward*," the whole will turn their eyes to the front and step off at a full pace.

The instructor will give his word "Forward" when he sees that the men are commencing the pace that will bring the front of the Squad perpendicular to the direction in which he intends it to move; this may be done at any degree of the circle.

In like manner, the squad will be taught to change direction to the left.

S. 30. The Diagonal March.

This march will first be taught commencing from the halt, after which the soldiers, when marching in line, will be practised in moving diagonally to either flank, by making a half-turn in the direction required.

Half Right-Turn. Slow—March.—

1. *From the Halt*.—On the word "*Turn*," the men will make a half-turn to the right; and on the word "*March*" each man will step off and move correctly in the diagonal direction, no longer keeping the touch.

The right-hand man will direct, and must therefore pay particular attention to his direction and pace. Each of the other men will glance towards the right, and will retain his relative position, keeping his right shoulder behind the left shoulder of the next man on that side.

Squad, Halt-Front.—On the word "*Halt*" the squad will halt; and on the word "Front," it will turn to its original front.

If the diagonal march has been properly performed, the squad, when halted and fronted, will be found to be in a line parallel to its original position.

Half Right-Turn, Front-Turn.—

2. *On the March.*—When the squad is marching to the front, and is required to move in a diagonal direction to the right, the word "Half Right-Turn" will be given, upon which the men will turn half-right and move diagonally in that direction, as described from the halt; when it is intended to resume the original direction, the word "Front-Turn" will be given, on which every man will turn to his front and move forward without checking his pace.

In like manner, the diagonal march will be practised to the left, from the halt and on the march.

The diagonal march will also be practised in quick and double time.

S. 31. Marching as in File.

Soldiers will first be taught to commence marching as in file, from the halt; after which they must be taught, when marching in line, to turn to either flank as in file.

Right (or Left)-Turn.—

1. *From the Halt.*—Soldiers, when standing as in file, must be instructed how to cover each other exactly. The head of the man immediately before each soldier, when he is correctly covered, will conceal the heads of all the others in his front.

The strictest observance of all the rules for marching is particularly necessary when marching as in file.

Slow-March.—On the word "*March*," the whole will step off together, at a full pace, and will so continue to step without increasing or diminishing the distance between each other. No looking down, or leaning back is to be allowed. The leader is to be directed to march straight forward on some distant objects, the remainder of the men covering correctly during the march.

Squad, Halt-Front—On the words "*Halt-Front*," the soldiers will halt and turn to their original front, and, if the marching had been properly performed, their dressing will be found correct.

Right-Turn.—

2. *On the March.*—On the word "*Turn*" the soldier will turn to the right, and move on as in file.

Front-Turn—The original direction is resumed by giving the word "*Front-Turn*," on which the soldier will turn to the front, and then move on steadily in line.

In like manner soldiers will be taught to turn to the left from line and march as in file, and, when marching as in file, to turn again to the front.

The rules laid down in **S. 22**, No. 2, regarding the foot on which the soldier is to turn, must be strictly observed in a squad in single rank.

Marching as in file will also be practised in quick time, but never in double time.

S. 32. WHEELING AS IN FILE.

Right (or Left)-Wheel, or Right-about (or Left-about)-Wheel.— The Squad, when marching as in file, will be taught to change its direction, by wheeling to the right or to the left, or to the right (or left) about. The leading man will move round a quarter, or half, of the circumference of a circle having a radius of four feet; the other men following on his footsteps in succession, without increasing or diminishing their distances from each other, or altering the cadence, but lengthening the pace a little with the outer foot, as they wheel.

S. 33. MEN MARCHING AS IN FILE, FORMING SQUAD,

At the Halt, Front Form-Squad.—1. *Forming to the Front at the Halt.*—When the squad, marching as in file to the right, is ordered to form to the front at the halt, the leading man will at once halt: the remainder will make a half-turn to the left, and form upon him as directed in **S. 28, No. 2.**

Eyes-Front.—Heads and eyes will be turned to the front.

Front Form-Squad.—2. *Forming to the Fronts in Quick Time.*— When a squad, marching as in file to the right, is ordered to form to the front, the leading man will mark time; the remainder will make a half-turn to the left, and form upon him, marking time, and taking up the dressing, as they arrive at their places.

Forward.—As soon as the squad is formed, the word *"Forward"* will be given.

On the March, Front, Form-Squad.—3. *Forming to the Front, in Double Time.*—When a squad, marching as in a file to the right, is ordered to form to the front on the march, the leading man will continue moving on; the remainder will make a half-turn to the left, double up to their places, and take up the quick time as they successively arrive in line with the leading man.

Rear Form-Squad. Forward or on the March, Rear Form-Squad.—4. *Forming to the Rear, in Quick, or Double, Time.*—The movement will proceed as described in No. 2 or No. 3; except that the men will make a half-turn to the right, and form on the right of the leading man.

Right Form—Squad. Eyes-Front.—5. *Forming to the Right,*—When marching as in file to the right, and ordered to form to the right, the leading man will wheel to the right, take two paces to his front, and halt; the remainder will form in succession on his left, and be dressed as they get into their places. The word *"Eyes Front"* will be given when the squad is formed.

Right-about Form-Squad. Eyes-Front.—6. *Forming to the Right-about.*—When a squad, marching as in file to the right, is ordered to form to the right-about, the leading man will wheel to the right-about, taking two paces to his front, and halt; the remainder will march on as in file, wheeling to the right on the spot where the leading man has wheeled, and forming successively on his left, looking to the flank of formation for their dressing. The words *"Eyes Front"* will be given when the squad is formed.

When marching as in file to the left, a squad will be formed to the front, or rear, or to the left, or left about, on the same principle as it is formed to the front, or rear, or to the right, or right about, when the right is leading.

A squad marching as in file will resume its original front by the words "Halt-Front," or "Front-Turn."

S. 34. THE SIDE OR CLOSING STEP.

The side or closing step will now be practised, the men judging the time as laid down in **S. 21**, No. 2. Care must be taken that the shoulders are kept square, and the paces made in a direct line to the flank.

SQUAD DRILL, IN TWO RANKS.

S. 37[2]. FORMATION OF SQUAD IN TWO RANKS.

The squad will now be formed for drill in two ranks, sized from flanks to centre. The men will take their places in succession, commencing from the flank on which they are ordered to form; each rear-rank man will be placed one pace of 30 inches from his front-rank man, measuring from heel to heel, and will cover him correctly, the two men thus placed forming a "File." When the squad consists of an uneven number of men, the third man from the left of the front rank will be a "Blank" (or incomplete) "File." The file on the left of the right half squad will always be the centre of the squad.

S. 38. DRESSING,

The front rank will dress as described in **S. 25.** The rear-rank men will continue looking to their front, and will cover and correct their distances as the front-rank men take up their dressing,

S. 39. MARCHING TO THE FRONT AND REAR.

A squad in two ranks will be practised in the marches, and variations of step, which have been taught in single rank.

1. Touch.—The front rank will touch as directed in **S. 24.**

2. Covering and Distance.—While marching in line, the men of the rank in rear must accurately preserve their covering and distances.

3. Blank File while retiring.—When the squad turns to the rear, a blank file, after turning about, will step up and occupy the vacant space in the rear rank. On turning to the front, he will resume his original place.

S. 40. TAKING OPEN ORDER.

Open-Order.—On the word. "*Order*," The flank men of the rear rank will step back two paces in slow time, and turn to the right,

March.—On the word "*March*." The flank men will front and raise the disengaged arm horizontally from the elbow, and the rear rank will step back two paces.

---

2. N. B. There are no Ss. 35 or 36 in original book.

Rear Rank Dress.—On the word *"Dress."* The rear rank will dress by the right. Care must be taken not to move the flank men when dressing the rear rank.

Eyes-Front.—On the word *"Front."* The rear rank men will turn their eyes to the front, and the flank men will drop their arms.

Close-Order. March.—On the word "March," the rear rank will take two paces to the front in slow time.

The squad, if drilling with arms, will always, be ordered to shoulder before taking open order.

S. 41. RIFLE EXERCISES.

These exercises will now be practised.

S. 42. CHANGING FRONT BY WHEELING, OR FILE FORMATION.

The front rank of the squad will wheel, or form (forward), from the halt, or wheel on the march, according to the instructions laid down in **S. 28** or **S. 29**. The rear-rank men, in wheeling, will follow their front rank men, keeping their proper distances, and covering; in forming, they will preserve their diagonal position.

S. 43. THE DIAGONAL MARCH.

The diagonal march will be practised in two ranks, in the manner described in **S. 30**. In addition to the instructions there given, the rear-rank men must be cautioned to preserve their relative positions with their front-rank men, in order that they may be found to cover correctly when they are halted and fronted.

S. 44. FILE MARCHING, WHEELING IN FILES, AND FILES FORMING SQUAD.

1. File Marching.—File marching will be practised as laid down in S. 31, care being taken that the rear-rank men dress correctly by their respective front-rank men.

2. Wheeling in Files.—Wheeling in files will be performed as laid down in **S. 32**; the outer rank must step rather longer during the wheel, especially with the outer foot.

If a squad is halted, or ordered to mark time, when only some of the files have wheeled into the new direction, the remainder should be taught to cover off, if required, by the diagonal march on the words *Rear files, Cover*: if the word *Front* is to follow the word *Halt*, the rear files need not be ordered to cover, but will remove to their places on the word *Dress*,

3. Files forming to the Front, or Rear.—The front rank men of a squad marching in files will form to the front, or rear, as laid down in **S. 33**. When forming to the front, the rear-rank man of the leading file will move into his place as soon as there is room for him; the other rear-rank men will preserve their relative positions with their front-rank men.

4. Files forming to the Right, or Left, or to the Right or Left about.—In these formations, the rear rank will form as described in **S. 33**; the front-rank men will move round their respective rear-rank men, and form successively in front of them.

A squad marching in files will resume its original front by the word *Halt-Front*, or *Front-Turn*.

S. 45. The Formation of Fours.

The squad should now be made up to eight or nine files, and numbered from right to left; and it must be explained to the men that odd numbers are right files, and even numbers left files. But in order that the left four may always be complete, when there happens to be an odd number on the left of the squad, the left file but one, although an even number, will be a right file, and the left file of the squad, though an odd number, will act as a left file: in this case the third file from the left, being a right file without a left file, will be called an "Odd File," whether it is complete or blank.

The four men composing a right and left file will be considered as comrades in the field, and will act together, not only in forming fours, but on other occasions; they should therefore take notice of each other when they are told off.

Fours.—

1. *At the Halt.*—On the word "Fours." The rear rank will step back one pace of 18 inches.

Deep.—On the word "*Deep.*" The left files will take a pace of 24 inches to the rear with their left feet, and a pace of 24 inches to the right with their right feet.

Squad-Front.—On the word "*Front.*" The left files will move up in line with the right files, by taking a pace of 24 inches to the left with their left feet, and a pace of 24 inches to the front with their right feet; the rear-rank men will then close up to their proper distances from the front rank, by taking a pace of 18 inches to the front with their left feet.

Fours.—As already described.

About.—On the word *"About."* The squad will turn to the right about, and the left files will form on the right files, by taking a pace of 24 inches forward, with their right feet, and a pace of 24 inches to the left with their left feet.

Squad-Front.—On the word *"Front"* The squad will turn to the right about, and re-form two deep as already described.

Fours.—As already described.

Right.—On the word *"Right,"* The squad will turn to the right, and the left files will form on the right of the right files, by taking one pace of 24 inches to the right with their right feet, and one pace of 24 inches to the front with their left feet.

Squad-Front.—On the word *"Front."* The squad will turn to the left, and re-form two deep as already described.

Fours.—As already described.

Left.—On the word *"Left."* the squad will turn to the left, and the left files will form on the left of the right files, by taking one pace of 24 inches to the left with their left feet, and one pace of 24 inches to the rear with their right feet.

Squad-Front.—On the word *"Front."* The squad will turn to the right, and re-form two deep as already described.

2. *On the March.*—A squad on the march will be taught to form fours on the words "Fours-Deep," "Fours-Right," and "Fours-Left;" in these formations the left files will move precisely as when forming from the halt, the right files marking time two paces to enable them to do so.

On the word "Fours," the rank in rear will step short two paces; if the word "Deep" follows, the right files will mark time two paces, while the left files move to their places in fours; if the word "Right," or "Left," follows, the squad will first turn in the direction ordered, after which the right files will mark time two paces while the left files move to their places.

When moving to a flank in fours, on the words "Half-Right (or "Left")—Turn," each man will-make a half turn to the ordered flank, as in the diagonal march. When moving, diagonally or direct, to a flank in fours, on the words "Front (or "Rear") Turn," the men will turn as ordered, and then form two deep without further word of command, by the right files marking time two paces, while the left files get into their places, and the rank in rear regains its distance.

A squad moving to the front or rear, or to a flank in fours, may be ordered to "Form Two-deep," on which the left files will fall back, or step up, into their places in file, and the rear rank will close on the front rank, the right files marking time two paces.

A squad moving to a flank in file may be ordered to form "Fours-deep." On the word "Fours," the rear rank will incline from the front rank by a lengthened step in the diagonal direction, and on the word "Deep," the left files will move to their places in fours, the right files marking time two paces.

A squad moving to a flank by the diagonal march may be ordered to form fours to that flank; the men will at once turn into file, and then proceed as above described.

On the order to form fours while marking time, the rank in rear will step back a pace of 18 inches, on the word "Fours."

S. 46. FOURS WHEELING, AND FORMING SQUAD.

1. *Wheeling,*—A squad moving to a flank in fours will wheel to the right or left, or to the right (or left) about, in the same manner as it wheels in files; each four wheeling successively round the same point. When the word "Forward" is given during a wheel, the leading four will march straight forward in the direction in which it is then turned; the remainder following. If the squad is halted, or ordered to mark time, when only some of the fours have wheeled, the remainder will move as directed in **S. 44,** on the words "Rear Fours, Cover."

2. *Forming to the Front or Rear, to the Right (or Left), or to the Right (or Left) about.*—When a squad moving in fours to a flank is ordered to form to the front or rear, to the right (or left), or to the right (or left) about, it will at once form two deep, and then proceed as described in **Ss. 44** and **33.**

S. 47. A SQUAD FORMED IN FOURS CLOSING ON A FLANK, OR ON THE CENTRE, AND REFORMING TWO DEEP.

For the following practices 10 or 12 files are required.

On the Right (Left, or Centre), Close. Quick-March.—A squad having formed four deep, will be taught to close on the right, left, or centre: the four men on the named flank, or in the centre, standing fast, the remainder closing on them by the side step.

From the Right (Left, or Centre),Re-form Two Deep. Quick

March.—In re-forming two deep, on the word "*March*" the four men on the named flank, or in the centre, will stand fast, the remainder will open out from them by the side step, and the left files will move up to their places in line in succession as the intervals are opened for them; the rear-rank men will step up to their proper distances at the same time.

These formations will also he practised on the march. On the words "*On the Right (Left, or Centre) Close*," the flies on the named flank, or in the centre, will move steadily forward at the stepping short pace, the remainder will close on them by the diagonal march; when all are closed, on the word "*Forward*," the squad will move on with a full pace.

On the words "*From the Right (Left, or Centre) Re-form Two Deep*," the file on the named flank, or in the centre, will move steadily forward at the stepping short pace; the remainder will incline outwards by the diagonal march, and the left flies will move up into the intervals, in succession, as they are opened for them, the rear-rank men regaining their proper distances at the same time: when completely formed in two deep, on the word "*Forward*" the squad will move on with a full pace.

S. 48. Breaking off Files,

A squad will he taught to reduce its front by breaking off files; files will, as a general rule, be broken off from the directing flank. A certain number of files (suppose three) will be broken off, as follows:

Three Files on the Left, Right-Turn, Left-Wheel.—On the word "*Turn*," the named flies will turn to the right, and on the word "*Wheel*," they will wheel to the left, following the left flank of the remainder of the squad.

The front of the squad may be further reduced by any number of files (suppose two), as follows:

Two Files on the Left, Right-Turn. Left-Wheel.—

On the word "*Turn*," two more flies will turn to the right, and on the word "*Wheel*," will wheel to the left, following the left flank of the remainder of the squad; the three files already in rear will mark time, ,then .incline to their right by the diagonal march, and follow close in rear of the two files last broken off.

Any number of files (suppose three) that have been broken off may be again ordered to the front, as follows:—

Three Files to the Front.—On the word "*Front*," the named files will make a half turn, and double up into their places in line; the remaining files in the rear will incline to the left by the diagonal march, and step out till they cover the two files on the flank.

Two Files to the Front.—On the word "*Front*," the two remaining files will make a half turn to the left, and double up into line.

All the files may be brought to the front at once by the words "*Files to the Front.*"

Files that are broken off must lock up well, that they may not interfere with others who may be following them.

DISMISSING.

S. 54[3]. DISMISSING A SQUAD.

Right-Turn.—1. *Without Arms.*—On the word "*Turn*," the men will turn as directed.

Dismiss.—On the word "*Dismiss*" the front rank will take a side pace to the left, and the rear rank will take a side pace to the right; after a pause the squad will break off quietly.

---

3. N. B. There are no Ss. 49-53 in original book..

Setting-Up Drill

Position of the Soldier.

The exact squareness of the shoulders and body to the front is the first and great principle of the position of a soldier. The heels must be in line and closed; the knees straight; the toes turned out, so that the feet may form an angle of 45 degrees; the arms hanging easily from the shoulder; the hands open; thumb to the front, and close to the fore-finger; fingers lightly touching the thigh; the hips rather drawn back, and the breast advanced, but without constraint; the body straight and inclining forward, so that the weight of it may bear principally on the fore part of the feet; the head erect, but not thrown back; the chin slightly drawn in, and the eyes looking straight to the front.

Back Stick Exercise.

The Back Stick Exercise will be found efficacious in producing suppleness of the arms, as well as causing a free expansion of the chest, without fatiguing the body by the exertion of using them.

N.B.—The words in italics are the words of command and cautions to be given by the instructor.

The squad will fall in in single rank, standing at "Attention." (Fig. B, **Plate 1**.)

Caution.—"*Take Distance from the Left*" on the Word "*Right Close*," "*Quick March*." The left file will remain steady, the remainder close to the right, each man taking his distance by raising his left arm, turning the palm upwards, the finger-nails resting on, or lightly touching, the right shoulder of the man who stands on his left. The left arm must be kept straight, the head and eyes turned to the left, the shoulders square to the front. (Figs. A and B, **Plate 1**.)

"*Eyes Front*." Bring the arm smartly to the body. (**Plate 2**, Fig. A 1.)

PLATE 1

A   B
Taking Distance from the Left.

"*Right Turn.*" First Practice. (**Plate 2**)

"*One.*" Raise the left arm and seize the stick above the right shoulder. (Fig. A.)

"*Two.*" Bring the stick smartly across the body, allowing at the same time the right hand to slip along the stick to the right (Fig. B.)

"*Three.*" Raise the stick with the arms extended, the palm of the hands to the front, the fingers straight and pointing upwards. (Fig. C.)

"*Four.*" Extend the arms along the stick, and let it fall slowly behind the back to the full extent of the arms, expand the chest, raise the head, and incline the body forward. (Fig. D.)

"*Three.*" Resume the Third Motion by raising the stick over the

PLATE 2

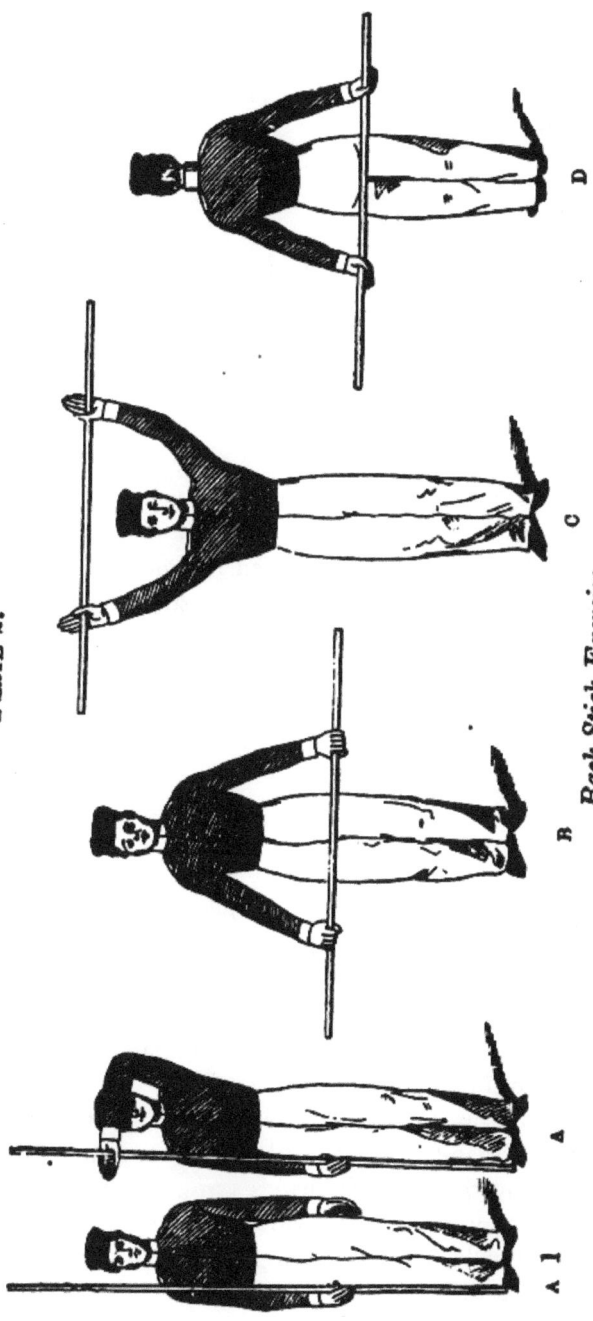

PLATE 2.

A 1 A B *Back Stick Exercise.* C D

head, as the stick is raised the hands must be brought a little nearer to each other.

"*Two*" Let the stick fall slowly down in front of the body, and resume the Second Motion,

"*One*" Raise the left arm, and place the stick upon the ground, resuming the First Motion.

In Quick Time,

"*One.*" Raise the left arm, seize the stick, and bring it smartly across the body. (Figs. A 2 and B.)

"*Two.*" Raise the stick over the head, and let it fall smartly behind the back, (from Second to Fourth Motion,) at the same instant bring it smartly over the head, in front of the body (from Fourth to Second Motion), (Figs. B, D.)

"*One.*" Resume the position of "Attention."

The Second Motion in Quick Time may be repeated several times, but not so long as to cause fatigue; about twelve times are sufficient, and then "Stand at Ease."

The chest must be freely expanded, the shoulders square, the head erect, and the body raised each time the stick is carried behind the back, or brought in front of the body.

On no account is the body to bend when performing any of the preceding motions. Instructors should check any tendency to such practice, which, if permitted, will considerably add to the unsteady gait, and shuffling walk, to which so many are liable, and which should be the constant aim of Instructors to eradicate or reform.

CLUB EXERCISE.

In order to supple the recruit, open his chest, and give freedom to the muscles, he must be exercised in the use of a pair of clubs (wooden), two feet in length, rounded and shaped to the hand, and their weight in the increasing proportion, of five, seven, and nine pounds each, in order that they may be used according to the strength of the men, and their progress in the exercise.

Standing in the "First Position of a Soldier," with the hands firmly grasping the handle of each club (pointing downward), the wrists tamed a little to the front; the hands, arms, and shoulders being perfectly easy and supple, without constraint, in their natural position.

The squad will be formed in single rank, with the clubs resting on the shoulders at the slope.

First Practice.
(Plate 3.)

Caution.—"*Take Distance from the Right*," on the word "*Left Close*," "*Quick March*." The right file will bring his clubs smartly down to the position of "Attention," and remain steady at the word "Quick March;" the remainder close to the left, each file taking distance from the right by raising the right arm extended, the club being six inches clear of the left shoulder of the file, who stands on his right. When that distance is taken, drop the clubs to the position of "Attention." (Figs. A and B.)

"*Stand at Ease*." Draw back the right foot; bend the left knee; let the toe of the club touch the ground, the palm of the hand resting on the top of the handle; the shoulders thrown back, and square to the front (Fig. B, **Plate 4.**)

"*Attention*." As before. (**Plate 3.**)

On the word "*One*." Turn the back of the right hand smartly to the front; bring the arm at the full extent slowly across and clear of the body; continue the motion in a circular direction, so that the end of the club clearing the left shoulder is carried over the head to the right shoulder, where it is held at the full extent of the right arm, the wrist to the front, the fingers straight. (Fig. A, **Plate 5.**)

On the word "*Two*." The same motions with the left arm and club; both arms remain extended above the shoulders. (Fig. C.)

On the word "*Three*." Let the arms fall slowly down until in line with the shoulders, the club held between the forefinger and thumb, the palm of the hands turned upwards, the fingers straight, the shoulders forced back, the chest expanded, the head erect, and the body inclining forward. (Fig D.)

On the word "*Four*" Lower the arms slowly and resume the position of "Attention."

The above motions may be performed alternately, and continuing them until the word "Steady" is given, when the position of "Attention" is resumed on the word "Four."

Caution.—Second Practice.

On the word "*One*." By combining the two First Motions, (A and B,) of the First Practice, raise both hands at the same time, and as the arms cross to the front, (having the right one uppermost,) continue

PLATE 3

A   B

Taking Distance from the Right.

PLATE 4

A   B
Attention.   *Stand at Ease.*

PLATE 5

(PLATE 5.)

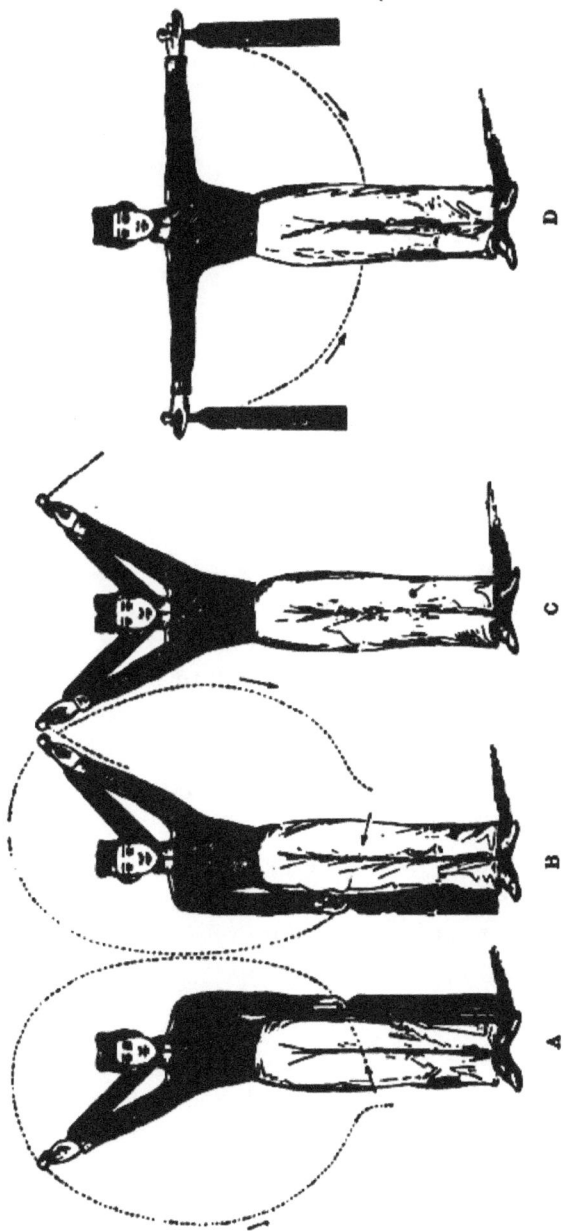

First and Second Practice.

the motion so as to raise them over the head, (Fig. C, First Practice,) and as they clear each other in doing so, (by the left arm rising over the right,) both hands must gradually fall to the rear, resuming the position of "Attention."

On the word "*Two.*" The same, commencing with the left arm uppermost.

On the word "*Three.*" The two preceding motions ("One" and "Two") alternately, without word of command, until ordered to discontinue the motions by the word "Four" being given, when the position of "Attention" will be resumed.

THIRD PRACTICE.
(Plate 6.)

On the word "One." Raise the hands up smartly in line with the chin, the hands grasping the handles of the clubs, the wrists touching; the knuckles to the front; and the arms extended, with the clubs perpendicular. (Fig. A.)

On the word "*Two.*" Force back the shoulders, and bring the hands smartly round in line with the shoulders, keeping the arms extended, the clubs perpendicular, and the wrists full to the front, (Fig. B.)

On the word "*Three.*" Turn the wrists upwards, and let the clubs fall slowly over to the rear, the handles resting firmly between the forefinger and thumb; the palm of the hands turned upwards, the fingers straight, the hands kept up in line with the shoulder, the head erect, and chest expanded. (Fig. C.)

On the word "*Four.*" Lower the arms to the sides, and resume the position of "Attention."

This last motion may be repeated, (*viz.*,) raising the arms from the position of "Attention" to the "Third Motion;" and as the arms are raised the body must be thrown forward, by raising the heels off the ground, and sinking them as the arms are lowered.

FOURTH PRACTICE
(Plate 7.)

"*Carry Clubs.*" Raise the clubs well balanced and perpendicular to the front of the breast, the hands about twelve inches apart, as high as the elbows, which must be well kept back, and the chest advanced. (Fig. A.)

On the word "*One.*" Raise the right-hand (the left remaining steady) into the hollow of the left shoulder, then by a quick turn of

PLATE 6

PLATE 6.

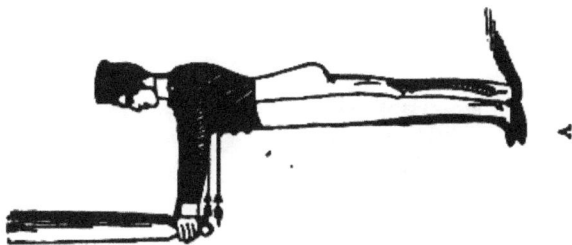

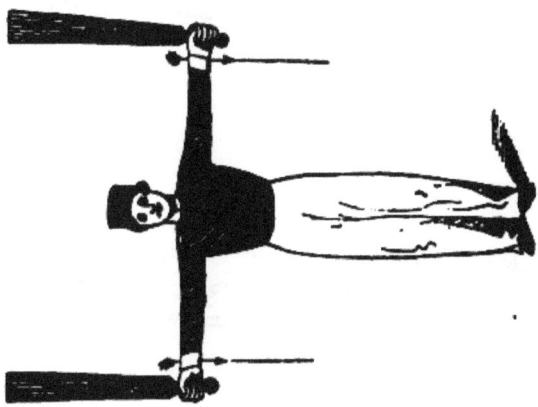

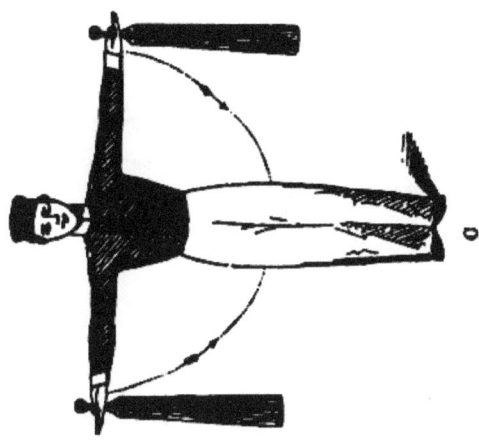

Third Practice.

A

B

C

PLATE 7

PLATE 7.

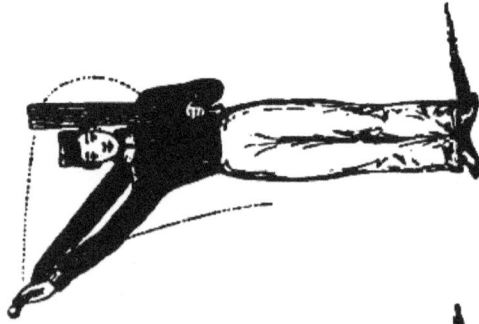

Fourth Practice.

the wrist and rise of the arm let the end of the club drop to the rear, continuing its motion round the head, when by the timely lowering of the elbow resume the position of Carry Clubs. (Fig. B.)

On the word "*Two.*" The same with the left arm, the right remaining steady to the front.

The two preceding motions, "One" and "Two" may be performed alternately, without word of command, until the word "Three" is given, when the hands return to the position of Carry Clubs.

"*Stand at Ease.*" Lower the Clubs. (**Plate 4,** Fig. B.)

FIFTH PRACTICE.
(**Plate 8.**)

On the word "*One.*" Raise the arms and clubs smartly up in front of, and in line with, the shoulders, the wrists turned a little upwards, and the arms straight (Fig. A.)

On the word "*Two*" Let the arms and clubs swing smartly to the rear, as far as possible without restraint, and bring the clubs smartly back to the First Motion. (Figs. A and B.)

As the clubs are swung to the rear the body must be raised and thrown well forward. The force of the clubs when passing the body to the rear will give an easy expansion to the chest, and a strengthening of the muscles of the arms.

"*Three,*" come to "Attention."

SIXTH PRACTICE. (SLOW TIME.)
(**Plate 9.**)

On the word "*One.*" Raise the right arm and club slowly to the front, continuing the motion until the arm and club are perpendicular with the shoulder; the wrist full to the front. (Fig. A)

On the word "*Two.*" Let the club fall slowly over to the rear, and resume the position of "Attention."

On the word "*One.*" The same with the left arm. (Fig. B.)

On the word "*Two.*" Resume the position of "Attention."

(IN QUICK TIME.)
(**Plate 10.**)

On the word "*One.*" Raise the right arm smartly, and continue to swing the club over the shoulder, performing a complete circle; the arms kept straight and shoulders square, until the word "Steady" is

PLATE 8

A   B

PLATE 9

A   B

PLATE 10

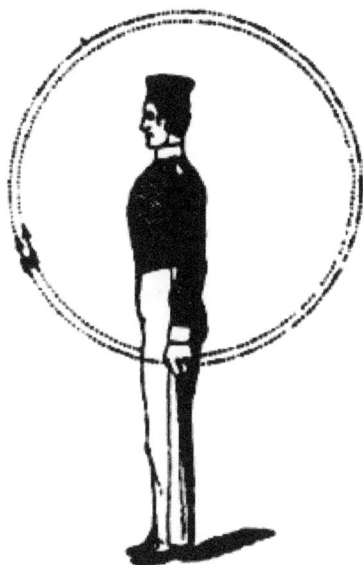

given, when the motion will be discontinued. (Fig. A.)

On the word "*Two.*" The same with the left arm.

On the word "*One.*" The same practice with both arms. On the word "Steady," discontinue.

Throughout the Club Drill, particularly those motions when the clubs are brought to the rear, great attention should be paid that the body be raised on the fore part of the feet, the shoulders kept square, and on no account be allowed to droop. The whole of the club exercise will be found perfect, as they combine free play of all the muscles of the body.

The object of all the movements in the club exercise is to supple the joints, and strengthen the muscles without constraining them by any forced positions: and great care should be taken that the position of "Attention" is strictly preserved, having the chest kept forward, and the head well raised from the lower part of the shoulder; in fact, the head rises, and the shoulders sink, in proportion as the bones of the chest are raised.

The easy and supple management of the clubs depends chiefly upon the timely turning of the wrists, and having the clubs well balanced, and by keeping an equal motion (either Slow or Quick, as may be ordered), without grasping them too tight; as otherwise the muscles become stiff, and the motions constrained.

Portions of the club exercise should be adopted daily by all young persons, more especially by those whose chests are deformed, which should be slowly and gradually expanded.

· The Back Stick or Club Exercise should never be used as a medium of punishment, by instructors, when a few are found dull (throughout the various stages of their drill). The recruit must be made to understand (when instructors have to send them occasionally to Club Exercise) that it forms a part of their drill, and is necessary to their training for the perfect soldier. Instructors must allow for the weak capacity of the recruit; and be patient, not rigorous, where endeavour and good will are apparent; for quickness is the result of much practice, and ought not at first to be expected.

EXTENSION MOTIONS.

These motions tend to expand the chest, raise the head, throw back the shoulders, and strengthen the muscles of the back. Men formed in squads with intervals will be turned a half-turn to the right, before commencing these practices.

CAUTION.—FIRST PRACTICE.
(Plate 11.)

One,—On the word "*One*" Bring the hands, at the full extent of the arms, to the front, close to the body, knuckles downwards, till the fingers meet at the points; then raise them in a circular direction over the head, the ends of the finders still touching and pointing downwards so as to touch the forage cap, thumbs pointing to the rear, elbows pressed back, shoulders kept down. (Figs. A and B.)

Two,—On the word "*Two*." Throw the hands up, extending the arms smartly upwards, palms of the hands inwards, then force them obliquely back, and gradually let them fall to the position of "Attention," elevating the neck and chest as much as possible. (Fig. C.)

Three.—On the word "*Three*." Raise the arms outwards from the sides without bending the elbow, pressing the shoulders back, until the hands meet above the head, palms to the front, fingers pointing upwards, thumbs locked, left thumb in front. (Fig. D.)

Four,—On the word "*Four*." Bend over until the hands touch the feet, keeping the arms straight; after a slight pause, raise the body gradually, bring the arms to the sides, and resume the position of "At-

Plate 11

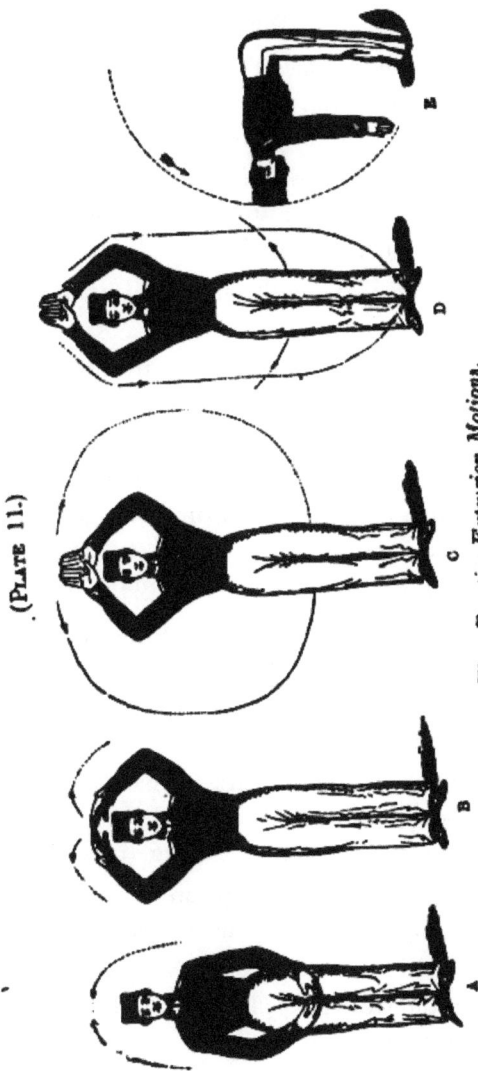

(Plate 11.)

First Practice Extension Motions.

tention." (Fig. E.)

The whole of these motions should be done slowly, so as to feel the exertion of the muscles throughout.

CAUTION.—SECOND PRACTICE.
(Plate 12.)

One,—On the word "*One.*" Raise the hands in. front of the body, at the full extent of the arms, and in line with the mouth, palms meeting, but without noise, thumbs close to the forefingers. (Fig. A.)

Two,—On the word "*Two*" Separate the hands smartly, throwing them well back, slanting downwards, at the same time raise the body on the fore part of the feet. (Fig. B.)

On the word "*One.*" Bring the arms forward to the position above described, and so on. (Fig. C.)

Three,—On the word "*Three,*" Smartly resume the position of "Attention."

In this practice, the Second Motion may be continued, without repeating the words *One, Two,* by giving the order, *continue the Motions;* the squad will then take the time from the right hand man. On the word "*Steady,*" the men will remain at the second position, and on the word "*Three*" they will resume the position of "Attention."

CAUTION.—THIRD PRACTICE.
(Plate 13.)

The squad will make a second half-turn to the right before commencing the Third Practice.

One.—On the word "*One.*" Raise the hands with the fists clenched, in front of the body, at the full extent of the arms, and in line with the mouth, thumbs upwards, fingers touching. (Fig. A.)

Two.—On the word "*Two.*" Separate the hands smartly, throwing the arms back in line with the shoulders, back of the hand downwards. (Fig. C, **Plate 15**)

Three.—On the word A "*Three.*" Swing the arms round as quickly as possible from front to rear.

Steady.—On the word "*Steady.*" Resume the second position. (Fig. C, **Plate 15.**)

Four.—On the word "*Four.*" Let the arms fell smartly to the position of "Attention."

PLATE 12

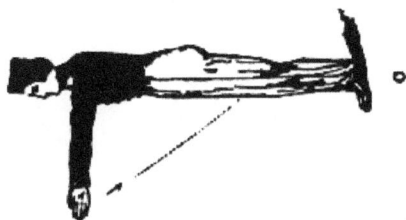

(PLATE 12.)

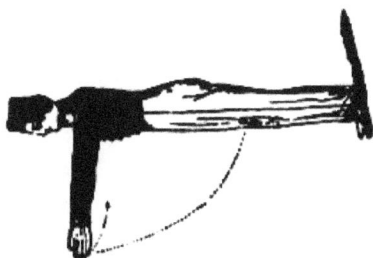

Second Practice.

PLATE 13

A

PLATE 14

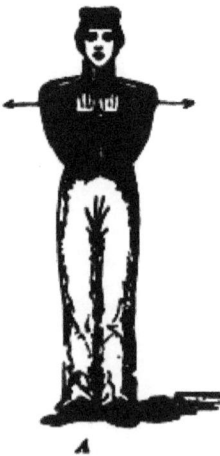

A

B

FOURTH PRACTICE.
(Plate 14.)

On the word "*One.*" Bring the shoulders to the front; raise the arms (from the elbows), the hands shut; the knuckles under the chin; the inner side of the hands touching, the elbows close and the hack of the hand to the front. (Fig. A.)

On the word "*Two.*" Separate the hands and press back the shoulders; keep the elbows close without constraint; the wrists full to the front; the knuckles in line with the shoulders. (Fig. B.)

On the word "*Three.*" Let the hands fall smartly, and resume the position of "Attention."

FIFTH PRACTICE.
(Plate 15.)

"*One*" Raise the elbows in line with the shoulders; the wrists bent; the backs of the hands touching the fingers shut, and the knuckles touching the top of the breast. (Fig. A.)

"*Two.*" Extend the arms smartly to the front, the fingers straight, the palms of the hands touching, and in line with the top of the breast; the thumbs close to the fore-finger. (Fig. B.)

"*Three.*" Separate the hands, and force the arms smartly back in line with the shoulders; the arms straight; the palms of the hands turned upwards, and the thumbs pointing to the rear. (Fig. C.)

"*Four.*" Lower the arms quickly to the position of "Attention."

SIXTH PRACTICE.
(Plate 16.)

"*One.*" Raise the hands to the full extent of the arms, above the shoulders; the palms to the front; the fingers straight, and pointing upwards; the thumbs pointing inwards. (Fig. A.)

"*Two.*" Bring the hands smartly down in front of the shoulders; the thumbs close to the forefinger; the shoulders pressed back, by forcing the elbows to the rear, and the body thrown forwards. (Fig. B.)

"Three." Let the hands fall smartly to the sides, and resume the position of "Attention."

SEVENTH PRACTICE.
(Plate 17.)

"*One.*" Raise the elbows in line with the shoulders, bend the wrists,

PLATE 15

(PLATE 15.)

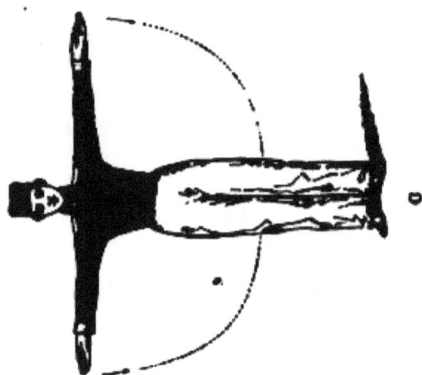

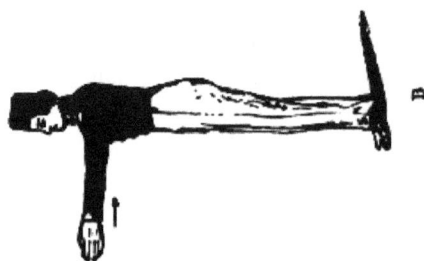

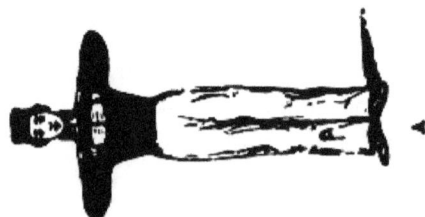

FIFTH PRACTICE.

A   B   C

PLATE 16

A

B

PLATE 17

A

and place the points of the fingers under the armpits. (Fig. A.)

"*Two*" Force the arms smartly down; press the elbows close to the body, the hands to the sides, with the palms turned out.

EIGHTH PRACTICE.
(Plate 18.)

"*One.*" Bring the hands, arms, and shoulders smartly round in front of the body, and at the full extent of the arms, the back of the hands touching the fingers straight. (Fig. A.)

"*Two.*" Raise the hands smartly up, in line with, and touching the upper part of, the breast, the elbow kept up in line with the shoulders. (Fig. B.)

"*Three.*" Press back the shoulders, and separate the hands smartly; bend the wrists; and place the point of the fingers on the top of the shoulders. (Fig. C.)

"*Four.*" Extend the arms smartly; the hands in line with the shoulders; the palms turned upwards; fingers straight; and thumbs pointing to the rear. (Fig. D.)

"*Five,*" Lower the arms quickly to the sides, and resume the position of "Attention."

The Third and Fourth Motion may be continued without repeating the words three and four.

NINTH PRACTICE.
(Plate 19.)

On the word "*One.*" Extend the arms smartly to the front; the palms of the hands touching, and in line with the breast; the fingers straight; and thumbs close. (Fig. A.)

On the word "*Two.*" Force the elbows smartly to the rear; expand the chest, and press back the shoulders; the palms of the hands to the front; the fingers pointing upwards; and close in front of the shoulders; the elbows close to the body. (Fig. B.)

On the word "*Three.*" Let the hands fell smartly to the sides, and resume the position of "Attention."

TENTH PRACTICE.
(Plate 20.)

On the word "*One*" Force the arms smartly up in front of the breast; the palms of the hands lightly touching; the fingers straight; and thumbs close. (Fig. A.)

PLATE 18

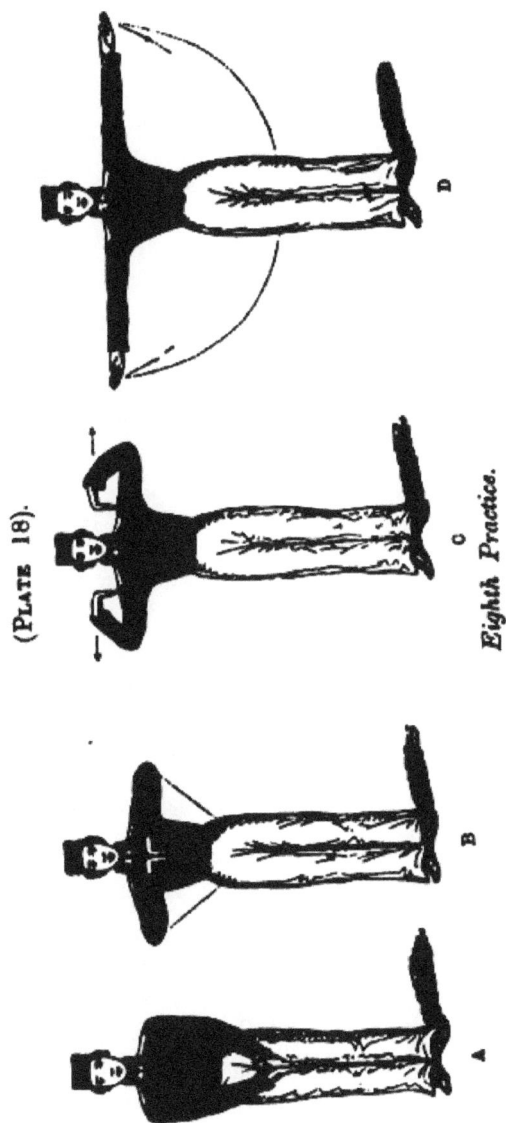

(PLATE 18).

Eighth Practice.

A

B

C

D

PLATE 19

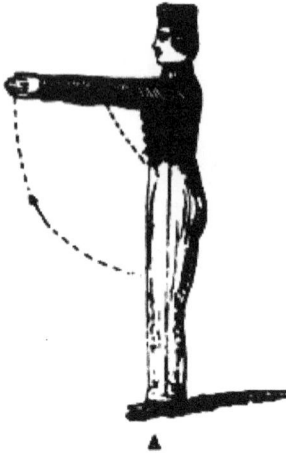

On the word "*Two.*" Draw the elbows close to the body; raise the hands smartly up close in front of the shoulders; the palms to the front, the fingers straight. (Fig. B.)

On the word "*Three.*" Extend the arms smartly; the hands in line with the shoulders; the palms turned upwards; the thumbs pointing to the rear. (Fig. C.)

On the word "*Four.*" Let the hands fall quickly down to the position of "Attention."

ELEVENTH PRACTICE.
(Plate 21.)

On the word "*One.*" Raise the arms smartly; bend the wrists; and place the points of the fingers on the top of the shoulders; the elbows being in line with the shoulders. (Fig. A.)

On the word "*Two.*" Extend the arms smartly above the shoulders, the palms of the hand to the front. (Fig. B.)

On the word "*Three.*" Bring the hands smartly down to the First Motion. (Fig. C.)

On the word "*Four.*" Let the hands fall smartly to the sides, and

PLATE 20

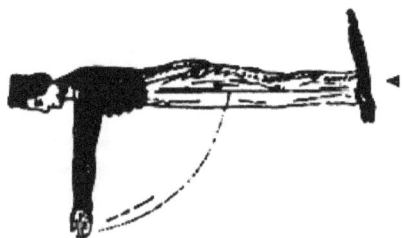

(PLATE 20).

Tenth Practice.

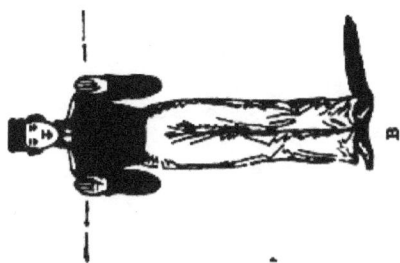

Plate 21

(Plate 21).

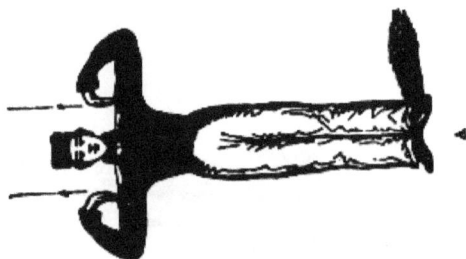

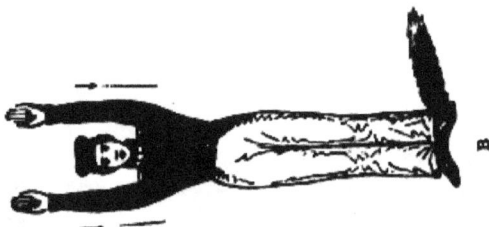

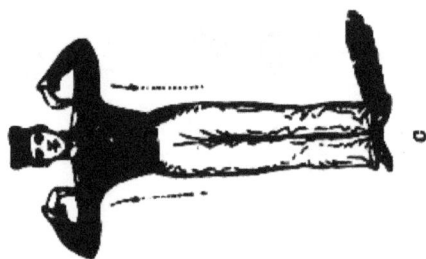

Eleventh Practice.

resume the position of "Attention."

The Extension Motions in slow and quick time, if judiciously managed, and the time and energy to be devoted to them, they might be made the most powerful means of counteracting the injurious effects arising from the disuse of particular muscles, and from the distorting and cramping positions to which persons of all classes, and various kinds of business are unavoidably subjected.

The whole of the preceding drill are intended to develop and strengthen the body and frame, and to secure a straight spine, and an erect and firm, but easy and graceful, carriage.

N.B.—The First Extension Motions—Second and Third Practice—is in accordance with practice laid down in the Field Exercise.

PREPARATORY MOVEMENTS FOR SWORD EXERCISE.

The following exercise is practised as a drill for the limbs only, and although it is not laid down in the Field Exercise for training the recruit, it will nevertheless be of the greatest benefit to the Rifle Volunteers and the youth of our schools.

It strengthens the muscles of the legs, as well as developing the upper parts of the body: and as it can be performed with the left shoulder and left foot to the front as well as with the right, it is of great importance that such exercise, so healthy and easily performed, should be constantly practised, as it brings into play all the muscles of the body at one time.

Caution.—First Position in Two Motions.
(Plate 22.)

On the word "*One*" Place the hands smartly behind the back, the left grasping the right just above the elbow, and the right supporting the left arm under the elbow.

On the word "*Two*," Make a *Half-turn Left*, by turning smartly on the left heel, bringing the right heel close and inside the left, the right foot pointing to the front, the feet at right angles, the whole weight of the body resting on the left leg. (Fig. 1.) The Second and Third Motions are here combined.

PLATE 22

A

In the above position the head will retain its position to the front

Caution.—Second Position in Two Motions.
(**Plate 23**.)

On the word "*One*" Bend the knees gradually, keeping them well apart; the body erect; feet firm on the ground (Fig. 1.)

On the word "*Two*." Keep the whole weight of the body on the left leg and place the right foot smartly about eighteen inches in front and in line with the left heel. (Fig. 2.)

In the Second Position, care should be taken that the left foot remains firm on the ground, and the right knee easy and flexible.

PLATE 23

1   2

Caution.—Balance Motions.

(Plate 24.)

On the word "*One.*" Move the right foot smartly about eight inches behind the left heel, the toe touching the ground, the heel perpendicular to it; keep the knees well apart. (Fig. 1.)

On the word "*Two.*" Raise the body slowly, by straightening the left knee. (Fig. 2.)

On the word "*Three.*" Bend the left knee, as in. First Motion. (Fig. 1.)

On the word "*Four.*" Advance the right foot, placing it firmly in front of the left, as in the Second Position.

On the word "*First Position.*" Bring the right foot smartly back to the left, and straighten both knees; the feet at right angles.

Caution.—Third Position in Two Motions.

(Plate 25.)

On the word "*One.*" Turn the right side smartly to the front, having the shoulder and knee perpendicular to the point of the right foot; the body kept perfectly erect, and resting firmly on the left foot. (Fig. 1.)

On the word "*Three.*" Step out smartly to front, double the distance as in the Second Position, the left knee firm and straight, the right heel in a line with the left, the body upright, and shoulders square to the left. (Fig. 2.)

Caution.—Second Extension Motions.

(Plate 26.)

On the word "*One.*" Bring the arms to the front of the body, with the hands closed, and the knuckles uppermost touching each other just below the lower button of the jacket; then raise them slowly until they come in a line with the upper part of the breast; keep the elbows well up, force back the shoulders by drawing the hands apart, then complete the motion by dropping the elbows, and smartly extending the arms and fingers in a diagonal line, the right hand as high as the head, the left in a line with the lower part of the left hip, the shoulders kept well down, and palms of the hands turned up. (Fig. 1.)

On the word "*Two.*" Raise the body by extending the right leg. (Fig. 2.)

On the word "*Three.*" Bend the right knee, and advance the body, resuming the First Motion.

Plate 24

1   2

Plate 25

1   2

PLATE 26

On the word "First Position." Spring up smartly, bringing the right heel close to the left, the arms behind the back, and resume the First Position.

On the word "Front." Smartly resume the position of "Attention."

In the foregoing, instructions having been explained by giving a separate word of command for each motion respectively, the same positions must now be gone through, naming only the word of command for the position required, so as to enable pupils to change quickly the positions without losing their balance. The words of command being used. *First, Second,* and *Third,*

Caution.—Positions.

On the word "*First.*" Raising the arms to the rear and the right heel to the front, come at once to the "First Position." (**Plate 22.**)

"*Second.*" Come to "Second Position." (**Plate 22**, Fig. 2.)

"*Third.*" Step out to "Third Position." (**Plate 25**, Fig. 2.)

"*First.*" Spring up to "First Position."

"*Second.*" "Second Position."

"*Third.*" "Third Position."

"*Second.*" "Second Position."

In all the above positions, each motion should be distinctly shown as before explained.

"*Single Attack.*" Raise the right foot about three inches, and beat it smartly on the ground.

"*Double Attack.*" Raise the right as before, and beat it twice on the ground, first with the heel, and then with the flat of the foot.

"*Advance.*" Move forward the right about six inches, and place it smartly on the ground, then bring up the left foot lightly about the same distance. "Single Attack" (as before).

"*Retire.*" Move the left foot lightly to the rear about six inches, the weight and balance of the body being and continuing to rest upon it: then move the right foot back the same distance, and place it smartly on the ground. "Double Attack" (as before).

"*Front.*" Resume the position of "Attention."

Instructors must take great care that the body is well balanced and rests on the left leg, thereby giving greater flexibility to the right leg in moving forward to gain distance on an adversary, or in retiring from

Plate 27

(Plate 27).

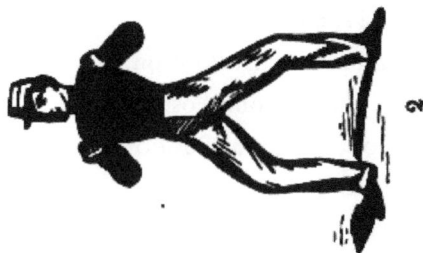

his reach. No precise length can be assigned in moving the right leg to the front in the "Third Position," as it depends upon the length and stride of the person, but it should not be beyond what may allow of his returning to the "First" or "Second Position" with quickness and perfect facility to himself.

The following combination of the First, Second, and Third Positions and "Second Extension Motions," as before explained, may now be practised.

Caution.—Positions and Longe by Numbers.

On the word "*One.*" Come to (as before explained) First Position, **Plate 27,** Fig. 1.

On the word "*Two.*" Come to Second Position (as before explained), bring the arms in front of the body, the hands closed and in line with the shoulders, which must be kept well down and back, the chest expanded, the head erect. **Plate 27**, Fig. 2.

On the word "*Three* ." Longe out to the "Third Position, extending the arms and fingers in a diagonal line, the right hand as high as the head, the left hand in line with the left hip, the palm of the hands turned up. **Plate 27**, Fig. 3.

"*One.*" Spring up to "First Position."

"*Two.*" Step out to "Second Position."

"*Three.*" Step out to "Third Position."

"*One.*" Spring up to "First Position."

After the squad has been sufficiently practised with the right side to the front, the instructor will give the word "Left," when the squad will turn quickly in that direction, the left side and left foot being to the front in "First Position."

"*Two.*" Step out to "Second Position," the hand closed and in line with the shoulders.

"*Three,*" Longe out to "Third Position;" the arms and hands extended, the left hand as high as the head, the right in line with the right hip; palms turned upwards.

"*One.*" Spring up to "First Position." "*Right,*" turn right side to front.

"*Front.*" Come to the position of "Attention."

The above practice may be repeated several times, care being taken not to cause fatigue by over exertion.

The practice may now be done in two motions.

On the word "*One.*" Come quickly to "First Position."

   „   „   „   "*Two.*" Longe out to "Third Position."

On the word "*One.*" Spring up to "First Position."

   „   „   „   "*Two.*" Longe out to "Third Position."

After several longes to Third Position, the word "Left" will be given, and the same practice continued with the left side to the front. Instructors should pay particular attention that at the longe, the arms and hands are properly extended, and that when the right side is to the front, the left knee must he well braced up, and the left foot kept firmly on the ground; the same attention that the right knee and foot is kept in the same manner when practising with the left side to the front.

On the word "*Right.*" The squad will turn quickly to the right.

"*Front.*" Turn to the front, and come to the position of "Attention."

As in all other exercises for the limbs, the foregoing practices may be gone through without separate words of command, the instructor giving the order to continue the motions by "Judging the Time."

FORMATION OF THE SQUAD FOR DRILL.

The right hand or left hand man being first placed, the remainder will fall in in line one after the other, closing lightly towards him, turning the elbow slightly outwards. Each man, when properly in line, should be able to feel his right or left hand man at the elbow; the body being preserved in the "First Position of a Soldier."

For the *whole* of the Setting-up Drill the squad will be formed with intervals as follows:—

Eyes Right.—On the Word "*Eyes Right*" The eyes will be directed to the right, the head being slightly turned in that direction.

Dress.—On the word "*Dress.*" Each soldier, except the right hand man, will extend his right arm, palm of the hand upwards, nails touching the shoulder of the man on his right; at the same time he will keep up his dressing in line by moving, with short, quick steps, till he is just able to distinguish the lower part of the face of the second man beyond him, care must be taken that he carries his body backward or forward with the feet, keeping his shoulders perfectly square in their original positions.

Eyes Front.—On the words "*Eyes Front.*" The head and eyes will be turned to the front, the arm dropped, and the position of the soldier resumed.

Dressing by the left will be practised in like manner.

Drilling a squad with intervals may be formed by directing the odd numbers, to take one pace forward, the even numbers to step back one pace.

If the squad is in two ranks, the *front rank* will be directed to take four paces forward, after which, odd numbers to take *two* paces forward and dress as above described.

STANDING AT EASE.

Caution.—Stand at Ease by Numbers.

One.—On the word "*One.*" Raise the arms from the elbows, left hand in front of the centre of the body, as high as the waist, palm upwards, the right hand as high as the right breast, palm to the left front; both thumbs separated from the fingers, and the elbows close to the sides.

Two.—On the word "*Two.*" Strike the palm of the right hand on that of the left, drop the arms to their full extent, keeping the hands together, and passing the right hand over the back of the left as they fall; at the same time draw back the right foot six inches, and slightly bend the left knee.

When the motions are completed, the arms must hang loosely and easily, the fingers pointing towards the ground, the right thumb lightly held between the thumb and palm of the left hand; the body must incline forward, the weight being on the right leg, and the whole attitude without constraint.

Squad-Attention,—On the word "*Attention.*" Spring up to the position described in Section 1, letting the arms fall, by the shortest way, to the sides.

JUDGING THE TIME.

Caution.—Stand at Ease, judging the time.

Stand at Ease.—On the word "*Ease.*" Go through the motions described in the standing at ease by numbers, distinctly but smartly, and without any pause between them.

Squad Attention.—As before.

If the command to *Stand at Ease* is followed by the word *Stand Easy*, the men will be permitted to move their limbs, but without quitting their ground, so that on coming to *Attention* no one shall have materially lost his dressing in line. If men are required to keep their dressing accurately, they should be cautioned not to move their left feet.

On the word squad being given to men Standing Easy, every soldier will at once assume the position of *Standing at Ease*.

A Catechetical Examination, for the Use of Instructors

The following brief catechetical examination of those who are intended to perform the duties of Drill Instructors will tend to ascertain their fitness for such an office; at all events it will prove that they have looked beyond the limits of the *Drill Book*, and had recourse to other works, or have learned what man is when uninstructed, what he has done, and what he is capable of doing when instructed:—

Q. What is the first duty you impress on the mind of the Recruit?

A. Obedience to superiors as administrators of law.

Q. How do you inculcate obedience?

A. By example.

Q. In what does the first part of military training consist'?

A. In improving the power of movement and action in the men who compose the squad.

Q. Is the perfection of that power important to all military operations?

A. Yes; it is the essential requisite through which all military operations when thoroughly understood are improved.

Q. What ought to be the principal object of the Drill Instructor?

A. To produce activity of bodily powers, united with mental sympathy.

Q. How is this unity of action to be effected?

A. By training the mind to a knowledge of all the movements of the body, and the intention of each movement, with its just value in accomplishing the end desired; thus joining theory to practice, and making one perfect by the help of the other.

Q. Is it possible to make the recruit understand the reason of *all* the movements he is taught?

A. No; actual service can only show him the reason of several evolutions which he cannot learn from his instructor. In this case, he must receive the rules and elements of learning implicitly, and trust to the future for the knowledge of those reasons on which they are grounded.

Q. Is it an advantage to the instructor to combine the theory of the drill with his daily lessons?

A. Yes; by so doing his physical labours are lessened, that is, he will have less trouble in correcting mistakes, as they will be generally corrected by his pupils, who will more readily see their errors.

Q. What advantage does the recruit derive from the knowledge of the theory of his exercise?

A. He will perceive with greater facility than he otherwise would that the whole of his exercise must be performed with a mechanical precision and correctness as to time; he will become also more interested in its execution the more he is informed of the principles upon which he, as one of a great body, is to act; and, when he is left to his own resources, he will be able to exercise his judgement in acting on those principles.

Q. If the union of mental with bodily power be the object which should principally influence you in preparing your Squad for the ranks, what means do you take to arrive at such an object?

A. My first step is to obtain a correct knowledge of the active powers of every recruit composing my squad; if I find that active powers predominate in one more than another; that is, if strong men are mingled with the weak, I make the former to be subordinate until the strength of the latter is increased, whereby the whole active powers of the squad may be combined.

Q. Will you not lose time by such a system?

A. By no means. I should if I acted otherwise. For example, if my squad be composed of fourteen, eight of whom are strong robust men, the other six growing lads; the whole possessing a teachable disposition, I give a certain task to be performed— perhaps with clubs. I insist upon the six performing their part equally with the eight; they strive to do so, but fail. If I still persist, they become weakened; they find they cannot keep up with their stronger brothers, and by over-exertion of the physical powers of the body, which produces exhaustion and fatigue, their whole system receives a shock, which incapacitates them from proceeding steadily with their drill; thus, that which was intended to strengthen and improve the men, is by a want of discernment in urging it too far, made to weaken and throw them back. If, on the other hand, I measure the power of my squad by the six, I lose no time, for the exertion is gradual, which tends to strengthen the physical energies of the weak, so that in due time they are able to equal the strong in all their future exercises.

Q. Do you arrange the men of your squad for instruction according to their external appearance, or intellectual powers?

A. According to both; for example, if the tallest men in my squad (who are on the flanks) are men possessing an inferior intellect, I have recourse to men of less stature, but superior mind, whom I place on the flanks, until their tall comrades are better acquainted with their work.

Q. Is it necessary to place men of quick mental powers on the flank of a squad?

A. Yes; and therefore I place such men there without reference to external appearance; if I act strictly according to external appearance, I lose many days, as all my attention must be directed to these men, for being on the flanks, they become the leaders of the squad, and a great deal depends upon their time of march, pace, and steadiness.

Q. Do you allow your pupils to retain the same place, or number in the squad, until finally dismissed to their duty?

A. No. If I allow them to retain the same place, or number in the squad, they would only learn the particular duty of that number, whereas, by changing them frequently, they learn the duties required from every file in a company.

Q. Are you able by this arrangement to judge better of the power and capabilities of each individual?

A. Yes; and after the men have thoroughly understood the exercise in all its parts, I arrange them according to their external appearance, thus combining strength with regularity of line. The utility of this arrangement consists in the order and union obtained in the disposal of the squad under all possible forms of movement.

Q. Is it necessary that you should understand the temper and former habits of each pupil?

A. Yes.

Q. By what means can you arrive at such a knowledge?

A. By keeping a careful watch upon him at drill, and in the barrack room; by observing the class of men with which he shows the greatest inclination to associate; the class of books he selects from the library; and, the manner in which he receives reproof; his former habits in civil life, and also his present inclinations, may thus be easily ascertained.

Q. What advantage do you gain by such a knowledge of your pupil's disposition?

A. That I may be able to use a proper discrimination in administering reproof. For example; the reproof for wilful errors would not be proper for those who err through a want of knowledge. Reproof also requires to be shaped according to the temper, of the individual to whom it is given; gentle, but firm, reproof generally attains its object.

Q. Having a right to command by the office which you hold, how do you deliver your commands?

A. By a mild, but firm, tone, which insures a cheerful obedience.

Q. Are there any other means besides your office which may assert your claim to obedience?

A. Yes; I show by cool, steady resolution, that I have a right to command, and to be obeyed; but, at the same time I adopt a gentleness in the manner of enforcing that right, and so make obedience cheerful, and soften as much as possible the consciousness of present inferiority.

Q. Is it a part of the instructor's duty to instil into the minds of his pupils sentiments, having a tendency to acts of bravery?

A. Yes; and as bravery is inspired by the force of example, I bring before them examples of perseverance in toil and difficulty, as a test of worthiness; and impress them with the one superior sentiment, to conquer or die on the field of battle, as an honourable sacrifice to the safety of their country.

Q. When can you have an opportunity of enforcing such a train of thought upon the minds of your pupils?

A. On wet days, which prevents the drill from being carried on outside, I endeavour to get possession of a spare barrack-room or shed, and there I impart to them a knowledge of their profession, and the sentiments it should inspire.

Q, How do you form your squad for drill?

A. That depends upon the kind of exercise they have to perform.

Q. Suppose you are called upon to put your squad through the Back Stick and Club Exercise?

A. In that case, I form my squad according to the instructions laid down in the manual of *Setting-up Drill*; and for any other kind of drill,—the authorized edition of the *Field Exercise.*

Q. What is the principal position of the soldier?

A. The exact squareness of the shoulders and body to the front is the first and great principle. The heels must be in a line, and closed; the knees straight; the toes turned out, so that the feet may form an angle of 45 degrees; the arms hanging easily from the shoulders, the hand open, thumb to the front and close to the forefinger, fingers lightly touching the thigh; the hips rather drawn back, and the breast advanced, but without constraint; the body straight and inclining forward, so that the weight of it may bear principally on the fore part of the feet; the head erect, but not thrown back, the chin slightly drawn in, and the eyes looking straight to the front.

Q. What is the principal intention of the First Position of a soldier?

A. That the recruit, on his first joining, may be well set up; or, in other words, placed upon his haunches in such a manner that all the joints of the body will bear equally and fairly upon each other.

Q. Why are the hips rather drawn in, and the breast advanced?

A. By the compression of the loins strength is thereby given to that part of the body; and when the breast is advanced, by throwing the shoulders back, an expansion takes place, which allows greater space for the action of the lungs.

Q. Which is the common centre of motion of the military figure?

A. The haunches; for all the movements performed in military tactics are referred to that part of the figure,

Q. Why is a just balance of the body necessary to the correct performance of movements?

A. Because no action of the body can possess power and ease of

148

movement, unless such movement be made justly and correctly upon the hips, as on a central pivot.

Q. How is the "First Position of a Soldier" to be maintained?

A. By an undeviating adherence to the rules laid down for it, and a frequent recourse to the motions and practice of the Setting-up drill: by the former, the position is maintained through the action and stability of the muscles, and the latter causes a successive action of the muscles, which thus relieve each other; and though the action be often repeated, it produces little or no fatigue.

Q. Does it tend to strengthen the muscles to continue their action beyond a limited time?

A. No; for if the action of the muscles be continued for a length of time without rest, fatigue ensues; and moreover, the contemplated act fails, under positions forced and constrained through fatigue.

Q. What is the first exercise performed by your pupils?

A. The Back Stick Exercise.

Q. What are the results arising from that exercise?

A. It is efficacious, in producing suppleness of the arms, as well as in causing a free expansion of the chest, without fatiguing the body by the exertion.

Q. What is the next exercise they perform?

A. The Club Exercise.

Q. What is the intention of the Club Exercise?

A. The object of all the movements of the clubs, as in the Back Stick Exercise, is to supple the joints and to strengthen the muscles, without constraining them by any forced positions.

Q. How is the ready and easy management of the clubs to be attained?

A. Their easy management depends chiefly upon the timely turning of the wrists, and maintaining an equal motion, having the clubs well balanced in the hands, and not grasping them too tight, lest the muscles become stiff, and the motions constrained.

Q. Do you ever cause your pupils to perform extra practice with the back stick, or clubs, or any other exercise, as a punishment?

A. Certainly not; however, when any are found to be dull, it becomes necessary that they should have extra practice, so as to enable them to keep pace with the more intelligent: in such a case, I inform them that it forms a part of their lessons, and is necessary to their training for the perfect soldier.

Q. What is the weight and length of the clubs?

A. Their weight is in the increasing proportion of five, seven, and nine pounds each, in order that they may be used according to the strength of the men and their progress in the exercise. They are two feet in length, rounded and shaped at the top for the hands.

Q. What exercise follows the Club and Back Stick?

A. The Extension Motions and other practices, laid down in the manual of *Setting-up Drill*.

Q. What are the Extension Motions and other practices intended to produce?

A. They cause an extension of size, or stature, elegance of figure, symmetry of limbs, ease and grace in movement. They can also be made, if judiciously managed, the most powerful means of counteracting the injurious effects arising from the disuse of particular muscles, and from the distressing and cramping positions to which persons of all classes, and various kinds of business, are unavoidably subjected.

B. B.

www.ingramcontent.com/pod-product-compliance
Lightning Source LLC
Chambersburg PA
CBHW021005090426

42738CB00007B/658